ESSENTIAL SKILLS IN MATHS

Answer Book

BOOK 1

Nelson

Graham Newman and Ron Bull

First published in 1996 by:
Thomas Nelson and Sons Ltd

Reprinted in 2002 by:
Nelson Thornes Ltd
Delta Place
27 Bath Road
CHELTENHAM
GL53 7TH
United Kingdom

05 / 10 9 8 7

A catalogue record for this book is available from the British Library

ISBN 0 17 431445 0

Printed and bound in China

National Curriculum coverage

Book	1	2	3	4	5
Levels	3–4	4–5	5–6	6–7	7–8

Contents

NUMBER

ALGEBRA

SHAPE, SPACE AND MEASURES

HANDLING DATA

Number

1 WORDS TO NUMBERS

Exercise 1A

1 226	**2** 411	**3** 831
4 306	**5** 517	**6** 213
7 644	**8** 4283	**9** 7376
10 5967	**11** 1915	**12** 5006
13 9119	**14** 250 000	**15** 70 620
16 15 705	**17** 7610	**18** 75 600
19 1 000 000	**20** 100 550	

Exercise 1B

1 647	**2** 835	**3** 954
4 110	**5** 488	**6** 346
7 612	**8** 5614	**9** 11 712
10 8600	**11** 17 076	**12** 74 500
13 20 650	**14** 21 000	**15** 5427
16 10 600	**17** 18 957	**18** 108 500
19 2 500 000	**20** 200 500	

Exercise 1C

1 5639	**2** 6320	**3** 2745
4 6008	**5** 3109	**6** 15 600
7 11 300	**8** 18 700	**9** 21 850
10 250 900	**11** 75 830	**12** 350 200
13 55 540	**14** 3 500 000	**15** 200 050
16 75 030	**17** 1 500 000	**18** 740 000
19 88 950	**20** 2 500 600	

Exercise 1D

1 6711	**2** 8220	**3** 6050
4 7320	**5** 1750	**6** 9712
7 12 500	**8** 25 000	**9** 15 740
10 67 838	**11** 47 600	**12** 205 700
13 75 950	**14** 39 100	**15** 500 000
16 1 500 000	**17** 4 500 000	**18** 1 650 000
19 25 000 000	**20** 405 500	

2 NUMBERS TO WORDS

Exercise 2A

1 Seven hundred and fifty
2 Four hundred and sixty-five
3 Two hundred and five
4 Four hundred and sixty-seven
5 Nine hundred and eighty-one
6 Six hundred and ninety
7 One thousand, four hundred and fifty
8 Three thousand, eight hundred and fifty
9 Two thousand, seven hundred and fifty-four
10 Nine thousand, eight hundred
11 Eighteen thousand, five hundred and five
12 Seventeen thousand and fifteen
13 Twelve thousand and thirteen
14 One hundred and sixty-five thousand, four hundred and fifty
15 One hundred and five thousand, eight hundred and seventy-five
16 One hundred thousand and fifty
17 One million
18 One hundred and seventy thousand, and ten
19 Eight hundred and fifty thousand, seven hundred and fifty
20 Five million, five hundred and six thousand, eight hundred and twenty

Exercise 2B

1 Three hundred and sixty-five
2 Five hundred and six
3 Two hundred and twelve
4 Five hundred and twenty-three
5 Nine hundred and fifteen
6 Seven hundred and fifty
7 Eight hundred and ninety-eight
8 Three thousand, five hundred and fifty
9 Five thousand, six hundred and seventy
10 Four thousand and fifty
11 Twenty-seven thousand, seven hundred and fifty
12 Thirty-six thousand, four hundred and fifty
13 Seventy-five thousand, four hundred and thirty
14 Three hundred and five thousand

15 Two hundred and twenty-five thousand, five hundred and seventy
16 Two million
17 Three million, five hundred thousand (three and a half million)
18 Forty-five million
19 Twenty thousand and twelve
20 Two million, seven hundred and eighty thousand, and fifty

Exercise 2C

1 Two thousand, five hundred
2 Two thousand, six hundred and fifty
3 Four thousand, five hundred and sixty-eight
4 Seven thousand, seven hundred and eight
5 One thousand and eighteen
6 Three thousand, nine hundred and five
7 Six thousand, five hundred and forty-one
8 Twenty-five thousand
9 Seventy-five thousand, seven hundred
10 Nineteen thousand, four hundred and seventy
11 One hundred thousand
12 One hundred and twenty-five thousand
13 Twenty-five thousand and fifty
14 One hundred and sixty-seven thousand, nine hundred
15 Two hundred and twenty five thousand, six hundred and nine
16 Twelve million
17 Seventy-five thousand
18 Three hundred and forty thousand, one hundred and fifty
19 One hundred and ten thousand, and eleven
20 One hundred and seven thousand, eight hundred and ninety-five

Exercise 2D

1 Three thousand, five hundred
2 Three thousand, four hundred and fifty
3 Four thousand, eight hundred and four
4 Two thousand and one
5 Five thousand and seventy
6 One thousand, one hundred and twelve
7 Eight thousand, nine hundred and seven
8 Twenty-five thousand
9 Thirty-four thousand, five hundred
10 Seventy-five thousand, six hundred
11 Thirty-seven thousand and forty-five
12 One hundred and eighty-five thousand
13 Six hundred and fifty thousand
14 One hundred thousand and forty
15 Twenty-five thousand, seven hundred and sixty-eight
16 Three million
17 Four million, five hundred thousand (four and a half million)
18 Three hundred and forty-seven thousand, eight hundred and ninety
19 Two hundred thousand, six hundred and eighty
20 One million, seven hundred thousand, five hundred

3 THE VALUE OF A GIVEN DIGIT WITHIN A NUMBER

Exercise 3A

1 600, 3
2 800, 40
3 300, 2
4 1000, 40
5 500, 2
6 9000, 100
7 500, 9
8 700, 40
9 8000, 1
10 20 000, 600
11 2000, 500
12 10 000, 20
13 9000, 4
14 90 000, 500
15 7000, 60
16 Three hundred, five
17 Six hundred, fifty
18 Seven hundred, nine
19 Two thousand, fifty
20 Eight hundred, five
21 Eight thousand, twenty
22 Six thousand, one hundred
23 Seven hundred, fifty
24 Four hundred, two
25 Eight thousand, seven hundred
26 One thousand, seven
27 Ten thousand, seven hundred
28 One thousand, nine hundred
29 Fifty thousand, seventy
30 Thirty thousand, seven thousand

Exercise 3B

1 500, 2
2 400, 30
3 300, 2
4 2000, 70
5 500, 40

6 3000, 5
7 9000, 300
8 600, 7
9 3000, 50
10 20 000, 70
11 30 000, 1000
12 20 000, 600
13 4000, 10
14 30 000, 500
15 2000, 5
16 Five hundred, one
17 Five hundred, sixty
18 Seven hundred, three
19 Two thousand, seven hundred
20 Six thousand, thirty
21 Four hundred, five
22 Three thousand, seven
23 Six thousand, fifty
24 Nine hundred, twenty
25 Eighty thousand, six hundred
26 Thirty thousand, two thousand
27 Forty thousand, one hundred
28 Seventy thousand, eighty
29 Four thousand, three hundred
30 Eighty thousand, one thousand

Exercise 3C

1 3000, 60
2 400, 8
3 5000, 20
4 20 000, 700
5 8000, 7
6 3000, 90
7 80 000, 300
8 90 000, 200
9 200 000, 100
10 1 000 000, 5000
11 400 000, 600
12 70 000 000, 400 000
13 500 000, 80
14 60 000, 7000
15 100 000, 7000
16 Two thousand, sixty
17 One hundred, five
18 Six thousand, fifty
19 Ten thousand, nine hundred
20 Four thousand, five
21 Ten thousand, seven hundred
22 Seventy thousand, eighty
23 Two hundred thousand, six thousand
24 Fifty thousand, six hundred
25 Eight million, one hundred thousand
26 Three hundred thousand, eight hundred
27 Twenty million, six thousand
28 Seven million, nine thousand
29 Six hundred thousand, five thousand
30 One million, nine thousand

Exercise 3D

1 5000, 2
2 200, 80
3 6000, 70
4 10 000, 300
5 20 000, 5000
6 70 000, 50
7 200 000, 400
8 30 000, 5000
9 30 000, 1
10 10 000, 600
11 1 000 000, 50 000
12 800 000, 900
13 20 000 000, 50 000
14 1 000 000, 200
15 10 000 000, 700 000
16 Six hundred, four
17 One thousand, seventy
18 Two hundred, nine
19 Ten thousand, forty
20 Thirty thousand, seven hundred
21 Five thousand, eighty
22 Two hundred thousand, three thousand
23 Three hundred thousand, five thousand
24 Seventy thousand, two thousand
25 Two million, seven thousand
26 Thirty million, forty thousand
27 Four million, seven hundred thousand
28 Ninety million, four hundred thousand
29 Twenty million, seven thousand
30 Six hundred thousand, fifty

4 ORDERING NUMBERS

Exercise 4A

1 45, 54, 78, 106
2 46, 87, 91, 127
3 79, 63, 34, 31
4 235, 193, 147, 89
5 86, 92, 143, 742
6 165, 531, 854, 861
7 365, 201, 109, 87
8 487, 468, 462, 395
9 301, 531, 813, 943

10 512, 633, 755, 811, 813
11 381, 298, 286, 173, 98
12 384, 361, 356, 354, 321
13 4820, 4653, 4561, 4210
14 4112, 4123, 4230, 4231
15 3519, 3521, 3531, 3541
16 2698, 2684, 2568, 2465, 754
17 9100, 8736, 8621, 8465, 7564
18 6012, 5866, 5864, 5799, 5763
19 997, 7556, 7986, 8644, 8975
20 8886, 7784, 6233, 5367, 4862

Exercise 4B

1 26, 55, 64, 86
2 64, 68, 78, 86
3 19, 23, 31, 32
4 145, 165, 172, 232
5 354, 269, 267, 176
6 832, 802, 799, 789
7 699, 730, 823, 864
8 389, 398, 400, 401
9 265, 258, 254, 245
10 866, 846, 832, 806, 789
11 500, 531, 574, 578, 590
12 795, 798, 868, 879, 986
13 1231, 1203, 1032, 1024
14 4006, 4022, 4060, 4566
15 986, 5897, 5987, 6488
16 5564, 4683, 4610, 3977, 3967
17 8977, 8779, 8634, 8577, 978
18 9002, 8001, 7982, 7888, 745
19 2487, 2556, 2665, 3110, 3112
20 5556, 5566, 5666, 6655, 6665

Exercise 4C

1 3120, 3211, 3222, 3224
2 1997, 2640, 2645, 3587
3 6543, 6897, 6898, 8856
4 685, 2642, 2645, 6851
5 7225, 7661, 7663, 7884
6 6862, 6733, 5001, 3986
7 6932, 6732, 6532, 3894
8 6656, 6565, 6555, 5656
9 14 561, 10 223, 6846, 5986
10 15 687, 99 866, 105 551, 114 364
11 801 633, 800 221, 88 330, 8654
12 668 566, 678 600, 687 800, 688 722
13 3 000 210, 3 000 009, 3 000 007, 3 000 001
14 500 000, 998 000, 1 250 000, 1 750 000
15 98 500, 99 000, 100 000, 101 500
16 487 000, 885 000, 4 999 999, 5 000 000
17 6 100 000, 6 000 500, 5 980 000, 5 875 000
18 4 050 000, 4 005 000, 498 999, 405 999
19 4 000 000, 3 650 500, 3 500 000, 2 900 000
20 1 100 000, 1 005 500, 1 001 500, 1 000 000

Exercise 4D

1 1320, 1254, 1025, 999
2 5100, 5011, 4908, 4900
3 2100, 2003, 1890, 1752
4 4654, 3865, 3666, 3265
5 8763, 8669, 7986, 7622
6 887, 8642, 8702, 8955
7 6223, 6284, 6332, 6731
8 4110, 4211, 5102, 5230
9 7850, 37 500, 38 560, 38 600
10 2 001 000, 2 000 400, 999 000, 200 500
11 1 080 000, 1 500 000, 1 550 000, 1 800 000
12 1 400 500, 1 100 000, 1 000 900, 1 000 500
13 78 540, 98 800, 99 500, 1 000 000
14 2 050 000, 2 000 500, 999 000, 199 500
15 999 000, 998 600, 988 800, 99 999
16 2 450 500, 2 450 000, 2 240 000, 245 000
17 49 900, 456 000, 465 200, 965 000
18 250 000, 2 000 000, 2 500 000, 2 650 500
19 65 550, 158 650, 223 000, 541 200, 684 100
20 12 350, 98 600, 123 500, 238 400, 850 000

5 CONVERTING POUNDS TO PENCE AND PENCE TO POUNDS

Exercise 5A

1 400p **2** 200p **3** 800p
4 600p **5** 1100p **6** 354p
7 571p **8** 375p **9** 475p
10 750p **11** 896p **12** 1216p
13 1735p **14** 1525p **15** 2560p
16 1438p **17** 1680p **18** 2868p
19 3062p **20** 2005p

Exercise 5B

1 500p **2** 900p **3** 700p
4 640p **5** 660p **6** 1785p
7 1306p **8** 1155p **9** 3978p
10 2131p **11** 3046p **12** 4587p
13 6384p **14** 10 030p **15** 12 040p
16 25 163p **17** 18 706p **18** 35 075p
19 46 644p **20** 20 432p

Exercise 5C

1 £0.57 **2** £0.65 **3** £0.89
4 £0.62 **5** £1.25 **6** £6.50
7 £1.75 **8** £2.04 **9** £2.95

10 £12.00	**11** £14.60	**12** £17.50
13 £10.54	**14** £8.90	**15** £38.75
16 £14.53	**17** £28.35	**18** £11.08
19 £19.45	**20** £40.06	

Exercise 5D

1 £0.84	**2** £0.65	**3** £0.89
4 £1.26	**5** £1.87	**6** £2.06
7 £1.56	**8** £14.25	**9** £28.96
10 £11.70	**11** £10.63	**12** £32.80
13 £14.45	**14** £265.23	**15** £123.23
16 £205.14	**17** £312.25	**18** £478.80
19 £354.00	**20** £624.05	

6 ADDITION AND SUBTRACTION OF MONEY

Exercise 6A

1 £2.99	**2** £6.87	**3** £4.51
4 £2.20	**5** £2.21	**6** £7.99
7 £6.41	**8** £8.20	**9** £2.28
10 £0.99	**11** £4.43	**12** £1.92
13 £2.09	**14** £10.00	**15** £1.16
16 £9.83	**17** £8.87	**18** £2.85
19 £2.28	**20** £8.48	**21** £0.51
22 £7.40	**23** £1.09	**24** £1.15
25 £8.07	**26** £3.99	**27** £7.61
28 £6.74	**29** £8.58	**30** £2.48

Exercise 6B

1 £6.14	**2** £5.91	**3** £1.93
4 £7.90	**5** £2.72	**6** £4.58
7 £8.11	**8** £3.29	**9** £3.17
10 £8.75	**11** £4.75	**12** £2.04
13 £3.45	**14** £9.27	**15** £4.78
16 £3.47	**17** £7.16	**18** £9.98
19 £10.00	**20** £3.10	**21** £2.52
22 £8.92	**23** £4.86	**24** £5.30
25 £10.70	**26** £1.75	**27** £1.89
28 £7.01	**29** £0.17	**30** £9.90

7 TOTALLING SUMS OF MONEY

Exercise 7A

1 £1.10	**2** £3.80	**3** £0.95
4 £5.18	**5** £4.00	**6** £1.69
7 £1.94	**8** £6.40	**9** £5.30
10 £8.30	**11** £26.80	**12** £43.00
13 £38.30	**14** £52.10	**15** £29.00
16 £109.00		

Exercise 7B

1 £2.85	**2** £1.86	**3** 4.70
4 £2.75	**5** £1.87	**6** £2.26
7 £3.95	**8** £13.74	**9** £11.10
10 £19.50	**11** £51.50	**12** £183.00
13 £73.60	**14** £49.50	**15** £13.20
16 £53.90		

Exercise 7C

1 £6.95	**2** £10.35	**3** £8.66
4 £9.00	**5** £9.19	**6** £17.88
7 £16.65	**8** £20.90	**9** £58.00
10 £13.65	**11** £39.04	**12** £160.00
13 £69.50	**14** £84.84	**15** £157.00
16 £150.50		

Exercise 7D

1 £4.85	**2** £14.50	**3** £15.24
4 £8.51	**5** £8.00	**6** £11.20
7 £10.29	**8** £46.40	**9** £78.50
10 £70.10	**11** £26.45	**12** £6.55
13 £174.50	**14** £56.90	**15** £89.50
16 £18.33		

8 MENTAL ARITHMETIC: ADDING AND SUBTRACTING

Exercise 8A

1 15	**2** 10	**3** 20	**4** 10
5 13	**6** 11	**7** 4	**8** 8
9 9	**10** 9	**11** 14	**12** 17
13 8	**14** 12	**15** 20	**16** 18
17 9	**18** 9	**19** 10	**20** 12
21 3	**22** 8	**23** 8	**24** 15
25 18	**26** 20	**27** 19	**28** 7
29 3	**30** 18	**31** 21	**32** 22
33 17	**34** 23	**35** 30	**36** 13
37 27	**38** 16	**39** 9	**40** 18

Exercise 8B

1 14	**2** 14	**3** 16	**4** 13
5 11	**6** 7	**7** 7	**8** 12
9 6	**10** 4	**11** 12	**12** 12
13 5	**14** 3	**15** 14	**16** 13
17 11	**18** 11	**19** 27	**20** 20
21 8	**22** 4	**23** 3	**24** 15
25 13	**26** 19	**27** 7	**28** 4
29 18	**30** 19	**31** 20	**32** 21
33 22	**34** 18	**35** 30	**36** 23
37 13	**38** 27	**39** 15	**40** 12

9 MENTAL ARITHMETIC: ADDING AND SUBTRACTING TWO 2-DIGIT NUMBERS

Exercise 9A

1 39 **2** 38 **3** 58 **4** 77
5 47 **6** 67 **7** 36 **8** 79
9 48 **10** 79 **11** 59 **12** 47
13 76 **14** 71 **15** 35 **16** 81
17 81 **18** 41 **19** 79 **20** 61
21 41 **22** 62 **23** 46 **24** 70
25 98 **26** 100 **27** 54 **28** 91
29 56 **30** 80 **31** 11 **32** 31
33 11 **34** 13 **35** 21 **36** 15
37 31 **38** 31 **39** 42 **40** 11

Exercise 9B

1 59 **2** 53 **3** 79 **4** 31
5 22 **6** 21 **7** 11 **8** 21
9 90 **10** 51 **11** 51 **12** 86
13 71 **14** 55 **15** 19 **16** 22
17 22 **18** 14 **19** 23 **20** 12
21 83 **22** 43 **23** 75 **24** 92
25 82 **26** 61 **27** 94 **28** 70
29 43 **30** 45 **31** 24 **32** 21
33 19 **34** 17 **35** 82 **36** 24
37 78 **38** 65 **39** 80 **40** 70

10 MENTAL ARITHMETIC: ADDING AND SUBTRACTING SEVERAL 1-DIGIT NUMBERS

Exercise 10A

1 12 **2** 20 **3** 22 **4** 18
5 20 **6** 18 **7** 13 **8** 19
9 13 **10** 22 **11** 9 **12** 8
13 8 **14** 4 **15** 2 **16** 18
17 14 **18** 10 **19** 5 **20** 11

Exercise 10B

1 14 **2** 13 **3** 13 **4** 20
5 16 **6** 18 **7** 16 **8** 17
9 17 **10** 13 **11** 3 **12** 12
13 7 **14** 6 **15** 5 **16** 17
17 20 **18** 5 **19** 10 **20** 3

Exercise 10C

1 14 **2** 20 **3** 24 **4** 22
5 21 **6** 22 **7** 27 **8** 23
9 30 **10** 25 **11** 14 **12** 8
13 8 **14** 10 **15** 2 **16** 16
17 2 **18** 10 **19** 2 **20** 11

Exercise 10D

1 18 **2** 17 **3** 22 **4** 16
5 28 **6** 18 **7** 26 **8** 20
9 24 **10** 28 **11** 15 **12** 13
13 8 **14** 3 **15** 5 **16** 12
17 8 **18** 9 **19** 9 **20** 7

11 PROBLEMS INVOLVING SIMPLE ADDITION AND SUBTRACTION WITHOUT A CALCULATOR

Exercise 11A

1 13 **2** 9 m **3** 13p
4 17 km **5** 14 **6** 6
7 1986 **8** 14 **9** 11
10 7 **11** 8 **12** 9°
13 9 litres **14** 15 **15** 12 cm
16 14 **17** 16 **18** 8
19 9 **20** 6

Exercise 11B

1 9 m **2** 13 km **3** 9
4 17 litres **5** 12° **6** 19
7 9 kg **8** 17 cm **9** 18 May
10 7 **11** 17 **12** 5
13 7 min **14** 13 **15** 12 litres
16 14 years **17** 7 **18** 19
19 9 **20** 15 days

Exercise 11C

1 24 **2** 8 litres **3** £20
4 9 litres **5** 27 hours **6** 15 min
7 26 **8** 7 **9** 23
10 £7 **11** 7 **12** 24
13 £7 **14** 21 **15** £24
16 12 **17** 5 kg **18** 5
19 18 **20** 19 days

Exercise 11D

1 24 kg **2** 16 **3** 15 years
4 20 hours **5** 11 **6** 20 pints
7 16 **8** 21 **9** 11.22
10 £12 **11** 24° **12** 13 cm
13 15 days **14** 12 **15** £21
16 24 km **17** 22 litres **18** 11 litres
19 23 **20** 9 hours

12 ADDING AND SUBTRACTING TWO 3-DIGIT NUMBERS WITHOUT A CALCULATOR

Exercise 12A

1 469 **2** 529 **3** 956 **4** 643
5 570 **6** 491 **7** 591 **8** 311
9 252 **10** 240 **11** 140 **12** 219
13 391 **14** 675 **15** 450 **16** 199
17 962 **18** 242 **19** 313 **20** 102

Exercise 12B

1 359 **2** 568 **3** 639 **4** 758
5 580 **6** 580 **7** 700 **8** 251
9 215 **10** 408 **11** 242 **12** 309
13 391 **14** 389 **15** 627 **16** 511
17 562 **18** 182 **19** 785 **20** 54

Exercise 12C

1 645 **2** 545 **3** 901 **4** 658
5 308 **6** 930 **7** 801 **8** 181
9 482 **10** 181 **11** 475 **12** 128
13 105 **14** 272 **15** 396 **16** 744
17 228 **18** 478 **19** 730 **20** 465

Exercise 12D

1 791 **2** 483 **3** 790 **4** 200
5 390 **6** 764 **7** 672 **8** 314
9 176 **10** 291 **11** 495 **12** 467
13 528 **14** 121 **15** 704 **16** 371
17 491 **18** 381 **19** 249 **20** 488

13 MULTIPLICATION TABLES

Exercise 13A

TEST 1

1 12 **2** 20 **3** 9 **4** 15 **5** 2
6 20 **7** 12 **8** 6 **9** 15 **10** 16
11 10 **12** 24 **13** 25 **14** 12 **15** 21
16 0 **17** 28 **18** 18 **19** 30 **20** 14

TEST 2

1 18 **2** 21 **3** 3 **4** 20 **5** 24
6 24 **7** 30 **8** 30 **9** 35 **10** 60
11 12 **12** 10 **13** 28 **14** 0 **15** 16
16 14 **17** 18 **18** 40 **19** 15 **20** 9

TEST 3

1 8 **2** 20 **3** 21 **4** 50 **5** 18
6 12 **7** 6 **8** 20 **9** 24 **10** 21
11 30 **12** 12 **13** 20 **14** 12 **15** 5
16 27 **17** 20 **18** 24 **19** 12 **20** 15

TEST 4

1 20 **2** 21 **3** 8 **4** 30 **5** 18
6 9 **7** 10 **8** 12 **9** 12 **10** 14
11 50 **12** 6 **13** 15 **14** 20 **15** 12
16 16 **17** 40 **18** 16 **19** 25 **20** 27

TEST 5

1 18 **2** 20 **3** 9 **4** 10 **5** 30
6 8 **7** 15 **8** 12 **9** 14 **10** 40
11 15 **12** 16 **13** 21 **14** 20 **15** 20
16 18 **17** 18 **18** 20 **19** 16 **20** 12

TEST 6

1 12 **2** 10 **3** 15 **4** 8 **5** 15
6 6 **7** 12 **8** 9 **9** 25 **10** 16
11 20 **12** 30 **13** 20 **14** 9 **15** 8
16 15 **17** 12 **18** 20 **19** 10 **20** 30

TEST 7

1 15 **2** 12 **3** 6 **4** 20 **5** 12
6 40 **7** 16 **8** 25 **9** 18 **10** 14
11 50 **12** 9 **13** 20 **14** 8 **15** 30
16 6 **17** 15 **18** 16 **19** 15 **20** 40

TEST 8

1 10 **2** 12 **3** 25 **4** 12 **5** 12
6 50 **7** 14 **8** 20 **9** 18 **10** 60
11 15 **12** 12 **13** 30 **14** 15 **15** 14
16 12 **17** 6 **18** 20 **19** 8 **20** 40

Exercise 13B

TEST 1

1 12 **2** 21 **3** 8 **4** 15 **5** 6
6 16 **7** 18 **8** 20 **9** 0 **10** 20
11 12 **12** 25 **13** 16 **14** 15 **15** 18
16 24 **17** 0 **18** 24 **19** 10 **20** 12

TEST 2

1 18 **2** 27 **3** 40 **4** 0 **5** 30
6 28 **7** 36 **8** 60 **9** 32 **10** 30
11 15 **12** 20 **13** 21 **14** 18 **15** 50
16 12 **17** 12 **18** 15 **19** 4 **20** 20

TEST 3

1 27 **2** 14 **3** 10 **4** 9 **5** 0
6 28 **7** 30 **8** 16 **9** 18 **10** 25
11 15 **12** 21 **13** 16 **14** 24 **15** 42
16 27 **17** 60 **18** 3 **19** 20 **20** 30

TEST 4

1 12 **2** 20 **3** 18 **4** 28 **5** 18
6 30 **7** 18 **8** 14 **9** 15 **10** 12
11 30 **12** 36 **13** 70 **14** 32 **15** 36
16 8 **17** 25 **18** 56 **19** 45 **20** 48

TEST 5

1 42	**2** 40	**3** 49	**4** 24	**5** 40
6 45	**7** 63	**8** 35	**9** 16	**10** 35
11 56	**12** 50	**13** 63	**14** 54	**15** 48
16 81	**17** 27	**18** 64	**19** 24	**20** 40

TEST 6

1 15	**2** 32	**3** 54	**4** 15	**5** 56
6 27	**7** 20	**8** 35	**9** 48	**10** 30
11 24	**12** 9	**13** 56	**14** 63	**15** 40
16 36	**17** 28	**18** 18	**19** 25	**20** 42

TEST 7

1 16	**2** 21	**3** 45	**4** 50	**5** 24
6 72	**7** 21	**8** 24	**9** 18	**10** 35
11 36	**12** 27	**13** 40	**14** 49	**15** 30
16 12	**17** 45	**18** 63	**19** 48	**20** 16

TEST 8

1 14	**2** 40	**3** 12	**4** 64	**5** 28
6 18	**7** 70	**8** 42	**9** 20	**10** 24
11 36	**12** 25	**13** 72	**14** 32	**15** 54
16 60	**17** 81	**18** 18	**19** 40	**20** 28

14 DIVISIBILITY

Exercise 14A

1 2, 5, 10	**2** 5	**3** 2
4 2, 5, 10	**5** 2	**6** 5
7 None	**8** 5	**9** 2
10 2, 5, 10	**11** 5	**12** None
13 2	**14** 2, 5, 10	**15** None
16 5	**17** 2, 5, 10	**18** None
19 None	**20** 2, 5, 10	

Exercise 14B

1 5	**2** 2, 5, 10	**3** 2
4 2	**5** 5	**6** 5
7 None	**8** 5	**9** 2, 5, 10
10 2, 5, 10	**11** 5	**12** None
13 5	**14** None	**15** 2, 5, 10
16 5	**17** 5	**18** 2
19 2	**20** 5	

Exercise 14C

1 2, 3, 6, 9	**2** 2, 3, 4, 5, 6, 8, 9, 10
3 5	**4** 2, 3, 4, 6, 9
5 2, 4, 8	**6** 2, 7
7 5	**8** 2, 3, 4, 5, 6, 9, 10
9 3	**10** 5, 7
11 2, 4, 7, 8	**12** 2
13 None	**14** 5
15 2, 4, 5, 8, 10	**16** 2, 3, 6, 9
17 2, 3, 6, 9	**18** None
19 2, 3, 4, 6	**20** 2, 3, 4, 6, 7, 8, 9

Exercise 14D

1 2, 3, 4, 6	**2** 3, 7, 9
3 2	**4** 2, 5, 10
5 2, 4, 8	**6** 2, 4
7 2, 3, 4, 5, 6, 9, 10	**8** None
9 3	**10** 2, 3, 6
11 2, 3, 4, 6	**12** 3, 9
13 2, 4, 8	**14** 7
15 2, 3, 4, 6	**16** 3, 5
17 2, 3, 4, 6, 8	**18** 3, 9
19 2, 3, 4, 6, 7, 8	**20** 2, 4

15 DIVISION: TIMES TABLES IN REVERSE

Exercise 15A

1 7	**2** 5	**3** 5	**4** 8	**5** 6
6 4	**7** 6	**8** 3	**9** 2	**10** 4
11 8	**12** 7	**13** 3	**14** 6	**15** 2
16 6	**17** 3	**18** 7	**19** 8	**20** 5
21 10	**22** 5	**23** 4	**24** 7	**25** 9
26 10	**27** 8	**28** 9	**29** 3	**30** 9

Exercise 15B

1 4	**2** 4	**3** 7	**4** 7	**5** 7
6 6	**7** 6	**8** 5	**9** 8	**10** 10
11 5	**12** 5	**13** 3	**14** 3	**15** 9
16 9	**17** 8	**18** 4	**19** 5	**20** 2
21 10	**22** 3	**23** 6	**24** 6	**25** 9
26 8	**27** 7	**28** 9	**29** 8	**30** 10

Exercise 15C

1 5	**2** 8	**3** 9	**4** 7	**5** 9
6 3	**7** 7	**8** 4	**9** 4	**10** 8
11 4	**12** 7	**13** 9	**14** 6	**15** 8
16 6	**17** 6	**18** 5	**19** 8	**20** 8
21 7	**22** 9	**23** 5	**24** 6	**25** 3
26 6	**27** 7	**28** 5	**29** 8	**30** 8

Exercise 15D

1 3	**2** 8	**3** 6	**4** 3	**5** 6
6 5	**7** 9	**8** 8	**9** 5	**10** 8
11 7	**12** 9	**13** 9	**14** 7	**15** 7
16 6	**17** 6	**18** 4	**19** 6	**20** 7
21 9	**22** 8	**23** 8	**24** 9	**25** 9
26 5	**27** 8	**28** 4	**29** 4	**30** 5

16 SIMPLE NUMBER FACTORS

Exercises 16A and 16C

1 2×3
2 2×6, 3×4
3 2×2
4 2×7
5 3×7
6 2×12, 3×8, 4×6
7 2×11
8 3×5
9 2×4
10 11×11
11 7×7
12 5×7
13 2×26, 4×13
14 2×20, 4×20, 5×16, 8×10
15 5×5
16 2×15, 3×10, 5×6
17 3×9
18 2×21, 3×14, 6×7
19 2×16, 4×8
20 2×10, 4×5
21 2×14, 4×7
22 2×28, 4×14, 7×8
23 3×15, 5×9
24 2×25, 5×10
25 3×21, 7×9
26 2×24, 3×16, 4×12, 6×8
27 2×20, 4×10, 5×8
28 2×36, 3×24, 4×18, 6×12, 8×9
29 2×27, 3×18, 6×9
30 2×30, 3×20, 4×15, 5×12, 6×10

Exercises 16B and 16D

1 2×5
2 2×9, 3×6
3 2×13
4 3×3
5 3×7
6 2×8, 4×4
7 5×17
8 3×13
9 5×7
10 2×19
11 2×23
12 3×17
13 3×33, 9×11
14 5×13
15 2×21, 3×14, 6×7
16 5×25
17 3×11
18 2×17
19 2×33, 3×22, 6×11
20 3×25, 5×15
21 2×49, 7×14
22 2×50, 4×25, 5×20, 10×10
23 2×45, 3×30, 5×18, 6×15, 9×10
24 2×100, 4×50, 5×40, 8×25, 10×20
25 2×18, 3×12, 4×9, 6×6
26 3×27, 9×9
27 2×42, 3×28, 4×21, 6×14, 7×12
28 2×22, 4×11
29 2×32, 4×16, 8×8
30 2×60, 3×40, 4×30, 5×24, 6×20, 8×15, 10×12

17 MULTIPLICATION AND DIVISION BY A SINGLE-DIGIT NUMBER WITHOUT A CALCULATOR

Exercise 17A

1	160	2	24	3	12	4	170
5	90	6	21	7	120	8	15
9	123	10	52	11	19	12	18
13	256	14	25	15	140	16	30
17	115	18	37	19	244	20	13

Exercise 17B

1	129	2	17	3	96	4	310
5	68	6	22	7	86	8	28
9	255	10	12	11	31	12	94
13	19	14	111	15	24	16	81
17	46	18	180	19	26	20	13

Exercise 17C

1	294	2	195	3	16	4	376
5	159	6	301	7	378	8	32
9	180	10	19	11	147	12	162
13	304	14	98	15	504	16	82
17	93	18	747	19	109	20	295

Exercise 17D

1	450	2	45	3	423	4	120
5	712	6	120	7	285	8	318
9	430	10	93	11	122	12	477
13	84	14	462	15	175	16	32
17	784	18	539	19	92	20	53

18 MULTIPLYING AND DIVIDING BY 10, 100 AND 1000

Exercise 18A

1 120	**2** 6500	**3** 320
4 60	**5** 45 300	**6** 560
7 745	**8** 23	**9** 2530
10 8500	**11** 94	**12** 350
13 17 000	**14** 597 000	**15** 9450
16 67	**17** 809	**18** 750
19 10 900	**20** 880	**21** 41 200
22 40	**23** 90	**24** 43
25 19 700	**26** 5000	**27** 69 000
28 84	**29** 550	**30** 9400

Exercise 18B

1 4100	**2** 760	**3** 52 000
4 450	**5** 72	**6** 18
7 4870	**8** 80 900	**9** 67 210
10 960	**11** 70	**12** 19 000
13 30 000	**14** 8700	**15** 596
16 71	**17** 820	**18** 2820
19 11 000	**20** 11 000	**21** 101
22 430	**23** 67	**24** 850
25 68 000	**26** 8600	**27** 3000
28 320	**29** 61	**30** 5700

REVISION

Exercise A

1 (a) Six hundred and seven
(b) Two thousand, five hundred and sixty-seven
(c) Fifty-two thousand, one hundred and twenty-five
(d) Two million, five thousand, six hundred and fifty

2 (a) 136 (b) 2527 (c) 40 612 (d) 207 050

3 (a) Eighty (b) Seven hundred
(c) Four thousand (d) One hundred thousand

4 (a) 45, 56, 67, 78, 95
(b) 191, 198, 236, 455, 623
(c) 701, 2403, 3746, 3994, 4026
(d) 2345, 2354, 2534, 3245, 3425

5 (a) £1.07 (b) £3.25 (c) £30.25 (d) £0.85

6 (a) 135p (b) 97p (c) 1067p (d) 3218p

7 (a) £9.13 (b) £1.49

8 (a) 20 (b) 5 (c) 15 (d) 9 (e) 16
(f) 15 (g) 23 (h) 31 (i) 42 (j) 37

9 (a) 39 (b) 13 (c) 60 (d) 27 (e) 91
(f) 16 (g) 85 (h) 12 (i) 98 (j) 27

10 (a) 5 (b) 7 (c) 9 (d) 11 (e) 5
(f) 3 (g) 11 (h) 2

11 (a) 863 (b) 219 (c) 802 (d) 328 (e) 710

12 (a) 18 (b) 28 (c) 18 (d) 40 (e) 56
(f) 24 (g) 42 (h) 35 (i) 48 (j) 45

13 (a) 2, 5 and 10 (b) 2 only

14 (a) 3 and 9 (b) 2, 4, 5, 7 and 10

15 (a) 8 (b) 5 (c) 5 (d) 6 (e) 3
(f) 8 (g) 6 (h) 7 (i) 5 (j) 9

16 (a) 2×10, 4×5 (b) 2×21, 3×14, 6×7
(c) 5×7 only
(d) 2×36, 3×24, 4×18, 6×12, 8×9
(e) 7×7 only

17 (a) 170 (b) 26 (c) 30 (d) 5000
(e) 37 (f) 2010 (g) 7100 (h) 60 400
(i) 900 (j) 870 000

Exercise AA

1 33 810
2 42 200 000
3 124
4 (a) 9731 (b) 1379
5 £1.48
6 £16.08
7 22
8 12 km
9 7°
10 9
11 23 litres
12 5 sheets
13 38
14 £13
15 379
16 79
17 Two of (1×36), 2×18, 3×12, 4×9, 6×6
18 (a) 13 458 (b) 13 455
19 1000
20 950 cm

19 DIVISION TO THE NEAREST WHOLE NUMBER WITH OR WITHOUT A CALCULATOR

Exercise 19A

1 4	**2** 10	**3** 11	**4** 5	**5** 9
6 8	**7** 7	**8** 21	**9** 7	**10** 5
11 9	**12** 17	**13** 6	**14** 16	**15** 5
16 13	**17** 8	**18** 7	**19** 8	**20** 19

Exercise 19B

1 4	**2** 5	**3** 3	**4** 3	**5** 3
6 22	**7** 15	**8** 6	**9** 7	**10** 7
11 9	**12** 16	**13** 5	**14** 6	**15** 16
16 13	**17** 12	**18** 6	**19** 4	**20** 6

Exercise 19C

1 24	**2** 38	**3** 33	**4** 85	**5** 62
6 17	**7** 127	**8** 62	**9** 16	**10** 54
11 81	**12** 44	**13** 17	**14** 84	**15** 84
16 62	**17** 123	**18** 21	**19** 39	**20** 60

Exercise 19D

1 134	**2** 53	**3** 30	**4** 117	**5** 23
6 34	**7** 43	**8** 85	**9** 46	**10** 34
11 59	**12** 94	**13** 77	**14** 61	**15** 124
16 112	**17** 20	**18** 81	**19** 30	**20** 97

20 APPROXIMATING TO THE NEAREST 10 OR 100

Exercise 20A

1 30	**2** 50	**3** 40	**4** 40
5 70	**6** 100	**7** 80	**8** 20
9 30	**10** 40	**11** 100	**12** 200
13 100	**14** 200	**15** 300	**16** 300
17 300	**18** 400	**19** 400	**20** 600

21 (a) 80 (b) 100
22 (a) 630 (b) 600
23 (a) 650 (b) 600
24 (a) 720 (b) 700
25 (a) 750 (b) 700
26 (a) 750 (b) 800
27 (a) 920 (b) 900
28 (a) 840 (b) 800
29 (a) 860 (b) 900
30 (a) 560 (b) 600

Exercise 20B

1 30	**2** 40	**3** 40	**4** 50
5 70	**6** 70	**7** 90	**8** 70
9 70	**10** 80	**11** 200	**12** 500
13 300	**14** 400	**15** 200	**16** 400
17 600	**18** 600	**19** 700	**20** 600

21 (a) 90 (b) 100
22 (a) 540 (b) 500
23 (a) 850 (b) 800
24 (a) 850 (b) 900
25 (a) 900 (b) 900
26 (a) 980 (b) 1000
27 (a) 830 (b) 800
28 (a) 770 (b) 800
29 (a) 450 (b) 500
30 (a) 350 (b) 400

21 ESTIMATING THE ANSWERS TO ADDITIONS AND SUBTRACTIONS

Exercise 21A

1 141	**2** 72	**3** 142	**4** 98
5 119	**6** 275	**7** 94	**8** 25
9 147	**10** 171	**11** 803	**12** 504
13 916	**14** 228	**15** 168	**16** 637
17 256	**18** 1396	**19** 697	**20** 222

Exercise 21B

1 101	**2** 28	**3** 116	**4** 66
5 92	**6** 195	**7** 136	**8** 130
9 128	**10** 105	**11** 700	**12** 424
13 773	**14** 251	**15** 389	**16** 1414
17 289	**18** 794	**19** 976	**20** 961

22 NEGATIVE NUMBERS IN CONTEXT

Exercise 22A

1 Fall, 3°C	**2** £80	**3** 27 BCE
4 –4°C	**5** 50 years	**6** 0°C
7 £24	**8** Rise, 7°C	**9** 12.03
10 –2°C	**11** 11 CE	**12** 5 min
13 £50	**14** Rise, 3°C	**15** 14 min
16 £25	**17** –10°C	**18** Late, 8 min
19 Rise, 4°C	**20** 5 CE	

Exercise 22B

1 28 BCE	**2** –6°C	**3** 10.25
4 Fall, 9°C	**5** £20	**6** 1 min 43 s
7 Fall, 9°C	**8** £31	**9** –3°C
10 40 BCE	**11** Fall, 1°C	**12** 7 min
13 5 BCE	**14** 5°C	**15** £7
16 1 min	**17** Rise, 4°C	**18** £65
19 2 BCE	**20** –11°C	

23 PROBLEMS INVOLVING MULTIPLICATION AND DIVISION WITH AND WITHOUT A CALCULATOR

Exercise 23A

1 £2.10	**2** £8	**3** 7
4 39	**5** 120	**6** 12
7 60 min	**8** £13	**9** £1.61
10 £90	**11** 6 cm	**12** 13
13 72	**14** 15	**15** 144
16 165 ml	**17** 10	**18** 60 cm
19 21	**20** £2.80	

Exercise 23B

1 £16	**2** £266	**3** 200
4 £6.08	**5** 9	**6** 630 km
7 12	**8** £2.88	**9** £3
10 8 mm	**11** 56 h	**12** 126 kg
13 20	**14** 128	**15** 14
16 256 mm	**17** 20	**18** 108
19 18	**20** 135 kg	

Exercise 23C

1 £6.25 2 49p 3 192
4 24 5 6 6 12 m
7 £13.93 8 19p 9 72 kg
10 8 11 30 12 91
13 30 14 £2.43 15 8
16 17 17 30 18 450 kg
19 9 20 60

Exercise 23D

1 1235 cm 2 £9.73 3 £0.65
4 128 km 5 £43.65 6 £1.31
7 8 8 £20.40 9 £1.60
10 £15 11 31 12 £4.75
13 108 14 9 m 15 36
16 25 17 9 18 £8.05
19 120 kg 20 7

24 PROBLEMS INVOLVING ADDITION, SUBTRACTION, MULTIPLICATION AND DIVISION WITH AND WITHOUT A CALCULATOR

Exercise 24A

1 21 2 90p 3 £2.25
4 £1.22 5 £5.50 6 £2.32
7 29 cm 8 18 9 450 g
10 £18 11 £10.35 12 36 m
13 50p 14 35 cm 15 £5.85
16 300 s 17 316 cm 18 96
19 £17 20 £31

Exercise 24B

1 49 cm 2 152 cm 3 £4.30
4 12 5 £3.43 6 £1.20
7 £17 8 48 cm 9 £51
10 £6 11 5 12 13
13 474 14 34 15 200
16 £16 17 112 18 £8
19 100 20 1909

Exercise 24C

1 341 2 98 3 168
4 720 min 5 80 g 6 184
7 32 cm 8 392 9 105 s
10 40 11 450 min 12 15 mm
13 775 g 14 175 ml 15 625 ml
16 480 ml 17 357 18 82
19 £2.70 20 152

Exercise 24D

1 72 cm 2 38 3 138
4 £1.30 5 £38 6 1000 ml
7 33 cm 8 £52.90 9 1800 g
10 175 g 11 £15 12 £2.76
13 819 14 £224 15 48°
16 15 17 108 18 £1.38
19 78 20 £1.40

25 RECOGNISING SIMPLE FRACTIONS

Exercise 25A

1 $\frac{1}{2}$ 2 $\frac{2}{3}$ 3 $\frac{1}{4}$ 4 $\frac{1}{3}$
5 $\frac{3}{4}$ 6 $\frac{1}{5}$ 7 $\frac{2}{5}$ 8 $\frac{1}{8}$
9 $\frac{3}{8}$ 10 $\frac{2}{7}$ 11 $\frac{4}{7}$ 12 $\frac{7}{32}$
13 $\frac{3}{32}$ 14 $\frac{1}{9}$ 15 $\frac{2}{9}$ 16 $\frac{4}{9}$
17 $\frac{5}{16}$ 18 $\frac{9}{16}$ 19 $\frac{1}{12}$ 20 $\frac{7}{12}$

Exercise 25B

1 $\frac{2}{3}$ 2 $\frac{1}{4}$ 3 $\frac{1}{2}$ 4 $\frac{1}{3}$
5 $\frac{7}{8}$ 6 $\frac{3}{5}$ 7 $\frac{3}{16}$ 8 $\frac{11}{16}$
9 $\frac{2}{9}$ 10 $\frac{4}{9}$ 11 $\frac{1}{8}$ 12 $\frac{5}{8}$
13 $\frac{1}{10}$ 14 $\frac{7}{10}$ 15 $\frac{2}{15}$ 16 $\frac{7}{15}$
17 $\frac{7}{20}$ 18 $\frac{11}{20}$

Exercise 25C

1 $\frac{1}{4}$ and $\frac{3}{4}$ 2 $\frac{1}{3}$ and $\frac{2}{3}$
3 $\frac{1}{8}$ and $\frac{7}{8}$ 4 $\frac{2}{5}$ and $\frac{3}{5}$
5 $\frac{7}{16}$ and $\frac{9}{16}$ 6 $\frac{2}{9}$ and $\frac{7}{9}$
7 $\frac{3}{8}$ and $\frac{5}{8}$ 8 $\frac{11}{24}$ and $\frac{13}{24}$
9 $\frac{3}{8}$ and $\frac{5}{8}$ 10 $\frac{4}{15}$ and $\frac{11}{15}$

Exercise 25D

1 $\frac{1}{3}$ and $\frac{2}{3}$ 2 $\frac{2}{5}$ and $\frac{3}{5}$
3 $\frac{1}{4}$ and $\frac{3}{4}$ 4 $\frac{7}{12}$ and $\frac{5}{12}$
5 $\frac{3}{7}$ and $\frac{4}{7}$ 6 $\frac{4}{9}$ and $\frac{5}{9}$
7 $\frac{7}{16}$ and $\frac{9}{16}$ 8 $\frac{9}{20}$ and $\frac{11}{20}$
9 $\frac{11}{24}$ and $\frac{13}{24}$ 10 $\frac{15}{32}$ and $\frac{17}{32}$

26 DRAWING SIMPLE FRACTIONS

Exercise 26A

1
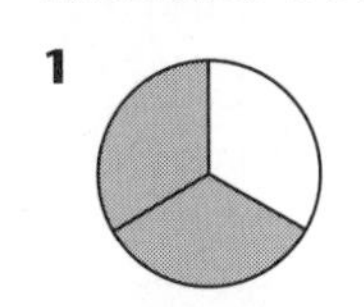

2

3
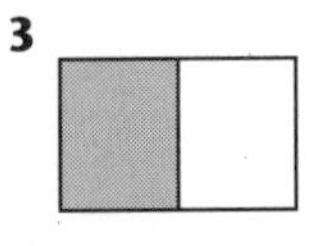

4
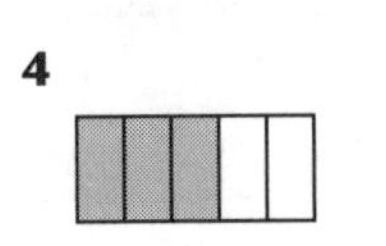

5
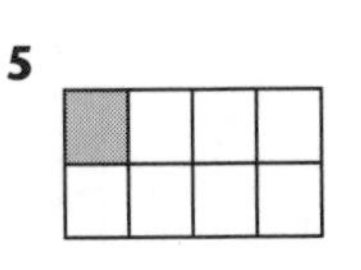

6
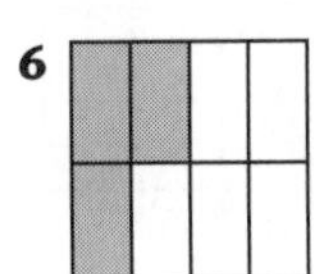

7
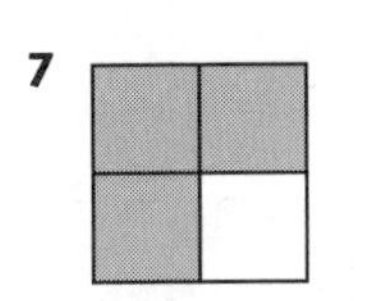

8
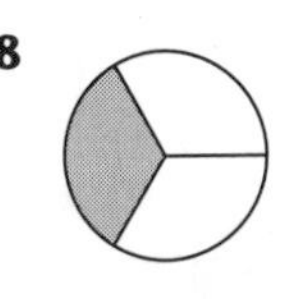

9
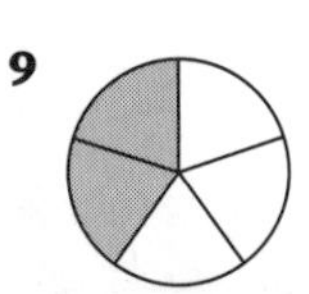

10
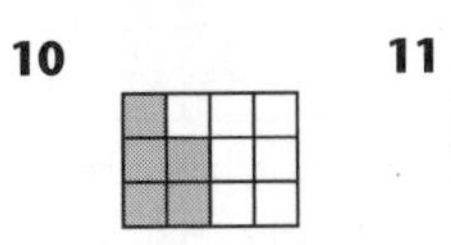

11
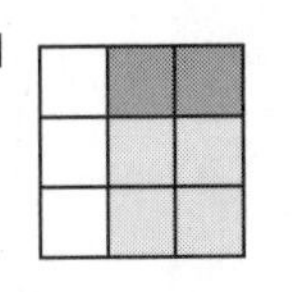

12
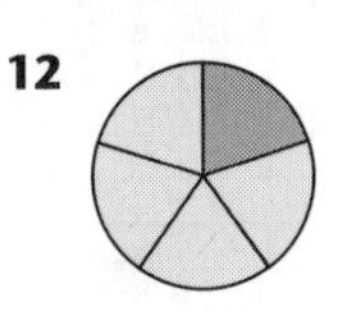

13
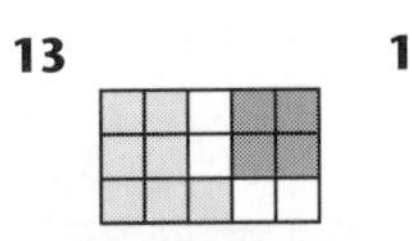

14
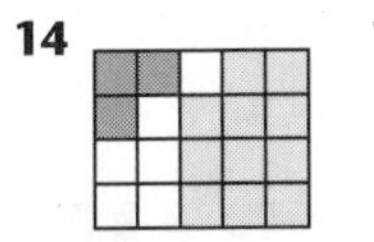

15
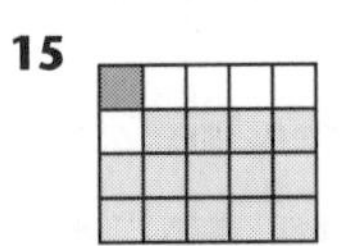

Exercise 26B

1

2
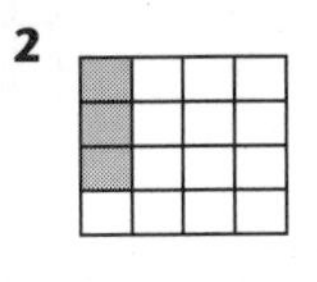

3
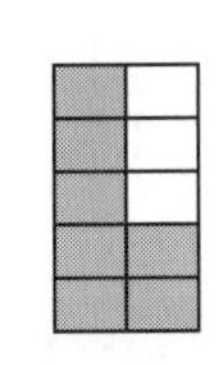

4
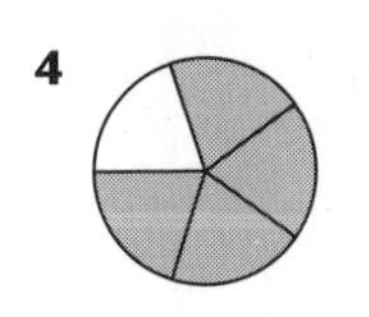

5
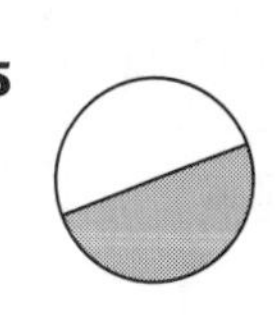

6
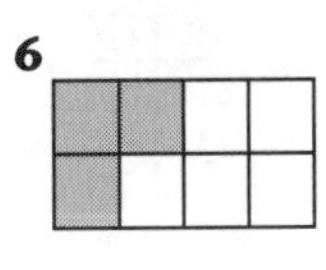

7
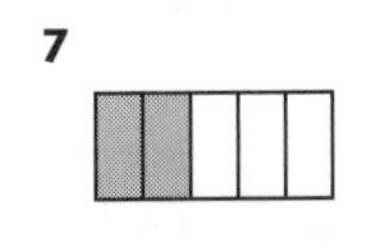

8
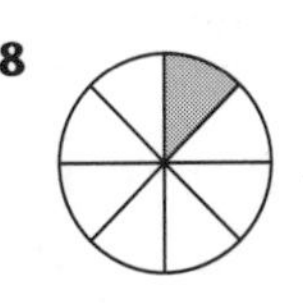

9
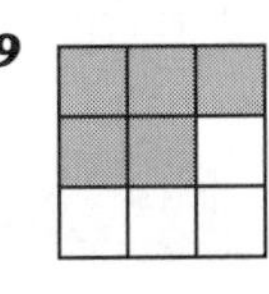

10
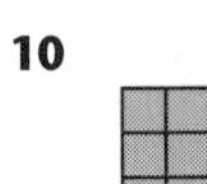
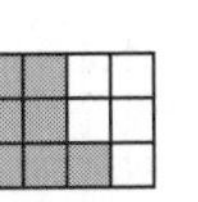

11
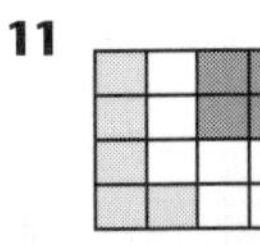

12
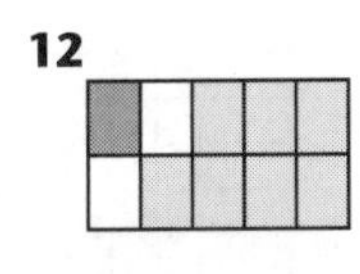

13

14
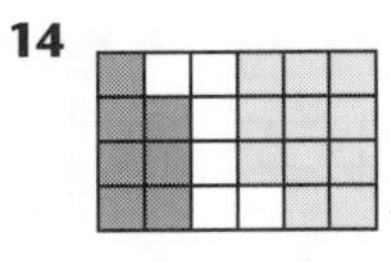

15
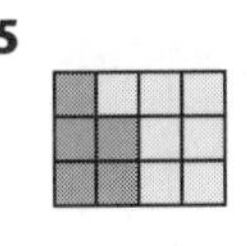

27 FRACTIONS OF QUANTITIES

Exercise 27A

1 5	**2** 28 kg	**3** £2
4 12 cm	**5** 5 km	**6** £12
7 £6	**8** 9 kg	**9** £8
10 £55	**11** 6 h	**12** £7
13 £8	**14** 50p	**15** 12
16 15 ml	**17** 12	**18** 2 min
19 5 g	**20** 25	

Exercise 27B

1 £2	**2** £20	**3** £2.25
4 20 km	**5** 2 km	**6** 16 h
7 £2	**8** 4p	**9** 16 km
10 19 ml	**11** 20 m	**12** 7 min
13 £2	**14** £2.50	**15** £4
16 5	**17** £4	**18** 80p
19 9 min	**20** 2 m	

28 RECOGNISING AND DRAWING SIMPLE PERCENTAGES

Exercise 28A

A 10%	**B** 40%	**C** 25%	**D** 7%
E 1%	**F** 96%	**G** 15%	**H** 80%
I 20%	**J** 75%	**K** 5%	**L** 90%
M 50%	**N** 30%	**O** 50%	**P** 10%
Q 60%	**R** 12%	**S** 20%	**T** 35%

Exercise 28B

A 30%	**B** 60%	**C** 85%	**D** 5%
E 25%	**F** 45%	**G** 90%	**H** 2%
I 50%	**J** 10%	**K** 4%	**L** 95%
M 35%	**N** 15%	**O** 1%	**P** 55%
Q 65%	**R** 3%	**S** 80%	**T** 9%

Exercise 28C

1
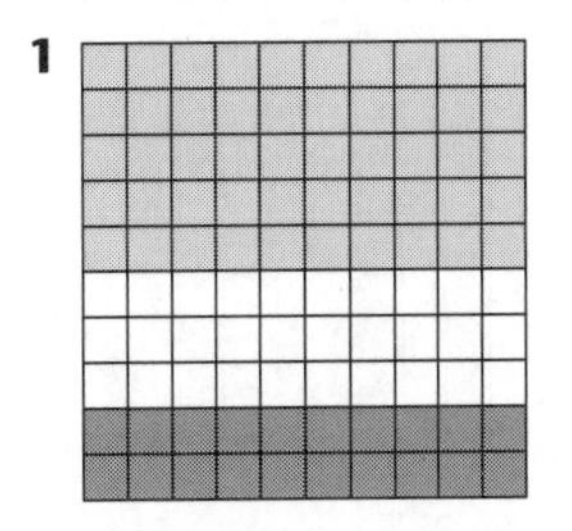

2
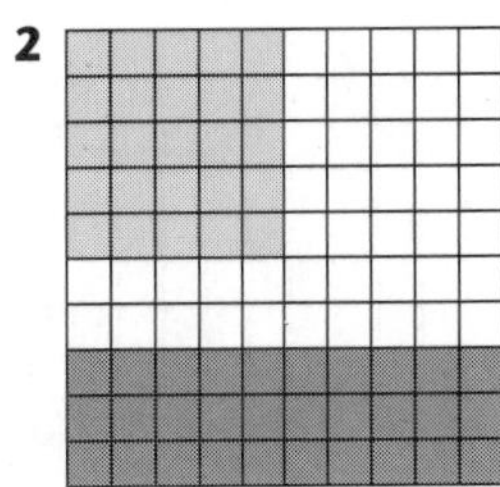

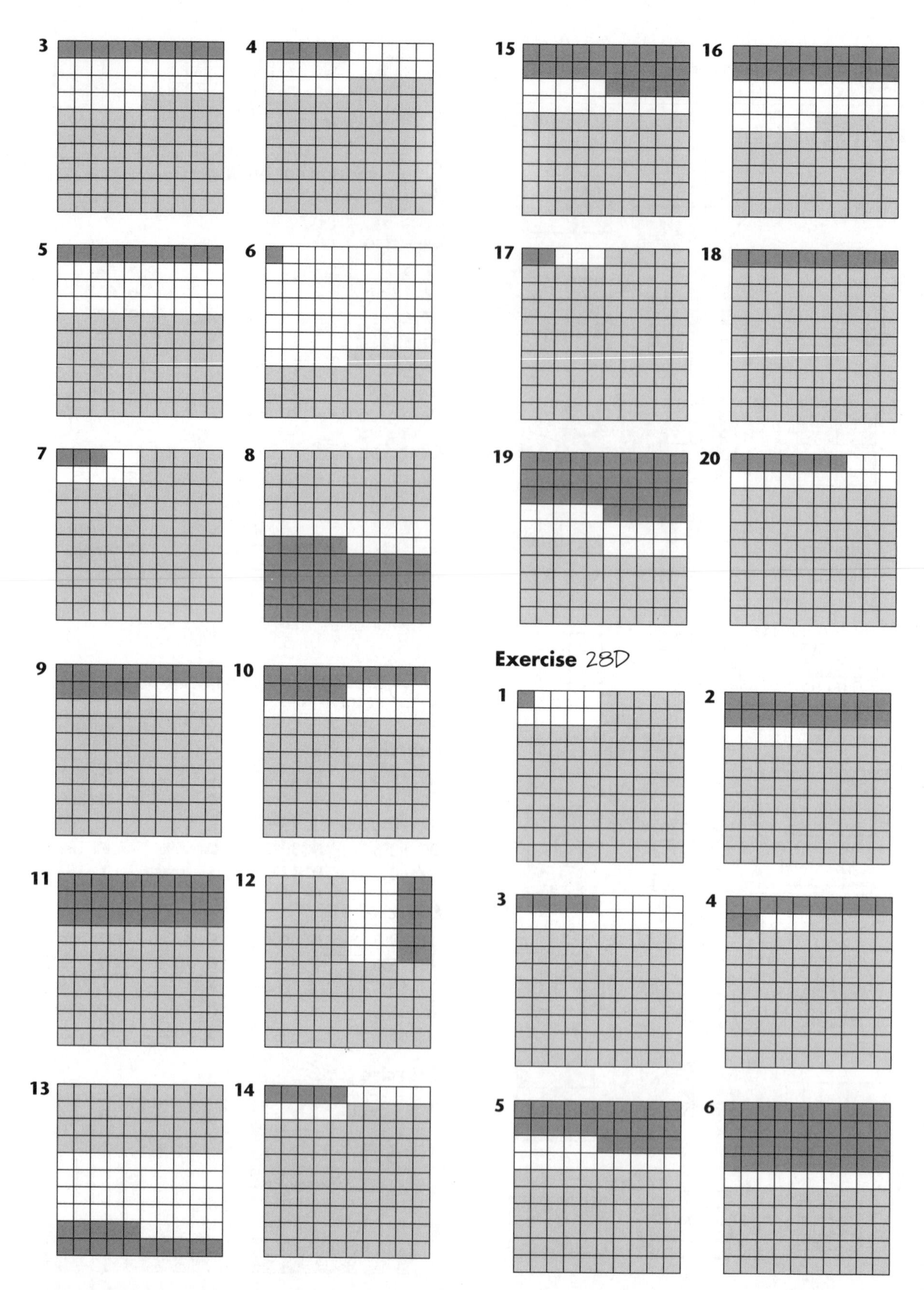

3
4
5
6
7
8
9
10
11
12
13
14
15
16
17
18
19
20
Exercise 28D
1
2
3
4
5
6

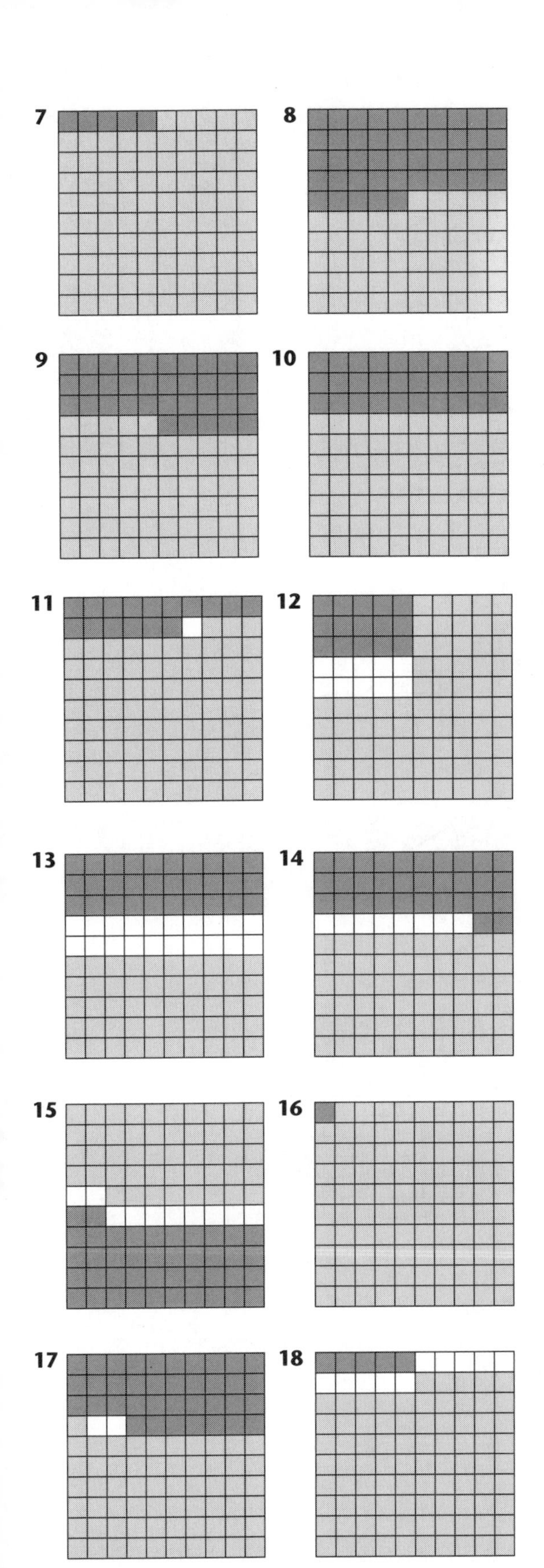

19

20

29 CONVERSION BETWEEN SIMPLE FRACTIONS AND PERCENTAGES

Exercise 29A

1 25%	**2** 75%	**3** 20%
4 90%	**5** 40%	**6** 1%
7 50%	**8** 30%	**9** 80%
10 97%	**11** $33\frac{1}{3}$%	**12** 10%
13 70%	**14** 60%	**15** 5%
16 83%	**17** 70%	**18** 29%
19 17%	**20** $66\frac{2}{3}$%	

Exercise 29B

1 60%	**2** $66\frac{2}{3}$%	**3** 20%
4 50%	**5** 35%	**6** 10%
7 90%	**8** 75%	**9** 1%
10 25%	**11** 80%	**12** 15%
13 55%	**14** 5%	**15** 73%
16 $33\frac{1}{3}$%	**17** 4%	**18** 95%
19 2%	**20** 83%	

Exercise 29C

1 $\frac{1}{5}$	**2** $\frac{1}{4}$	**3** $\frac{3}{10}$	**4** $\frac{1}{25}$
5 $\frac{3}{5}$	**6** $\frac{2}{5}$	**7** $\frac{13}{20}$	**8** $\frac{4}{5}$
9 $\frac{1}{50}$	**10** $\frac{1}{3}$	**11** $\frac{1}{2}$	**12** $\frac{1}{20}$
13 $\frac{61}{100}$	**14** $\frac{1}{10}$	**15** $\frac{3}{4}$	**16** $\frac{1}{100}$
17 $\frac{3}{20}$	**18** $\frac{99}{100}$	**19** $\frac{7}{10}$	**20** $\frac{11}{100}$

Exercise 29D

1 $\frac{4}{5}$	**2** $\frac{1}{20}$	**3** $\frac{3}{10}$	**4** $\frac{7}{10}$
5 $\frac{2}{5}$	**6** $\frac{1}{2}$	**7** $\frac{77}{100}$	**8** $\frac{3}{4}$
9 $\frac{9}{20}$	**10** $\frac{9}{100}$	**11** $\frac{1}{5}$	**12** $\frac{1}{3}$
13 $\frac{3}{50}$	**14** $\frac{19}{100}$	**15** $\frac{1}{4}$	**16** $\frac{1}{10}$
17 $\frac{17}{20}$	**18** $\frac{11}{20}$	**19** $\frac{3}{100}$	**20** $\frac{2}{3}$

30 SIMPLE PERCENTAGES OF SUMS OF MONEY

Exercise 30A

1 12p	**2** 10p	**3** 5p	**4** 28p
5 7p	**6** 6p	**7** 24p	**8** 48p
9 8p	**10** 20p	**11** 27p	**12** 8p
13 13p	**14** 12p	**15** 9p	**16** 18p
17 14p	**18** 16p	**19** 45p	**20** 15p
21 36p	**22** 32p	**23** 18p	**24** 13p
25 14p	**26** 45p	**27** 9p	**28** 45p
29 24p	**30** 28p	**31** 20p	**32** 64p
33 17p	**34** 50p	**35** 36p	**36** 30p
37 12p	**38** 30p	**39** 20p	**40** 77p

Exercise 30B

1 15p	**2** 15p	**3** 35p	**4** 30p
5 33p	**6** 16p	**7** 24p	**8** 26p
9 8p	**10** 13p	**11** 28p	**12** 42p
13 12p	**14** 18p	**15** 35p	**16** 27p
17 36p	**18** 70p	**19** 54p	**20** 40p
21 85p	**22** 32p	**23** 55p	**24** 24p
25 24p	**26** 30p	**27** 32p	**28** 18p
29 40p	**30** 40p	**31** 42p	**32** 44p
33 40p	**34** 63p	**35** 17p	**36** 56p
37 30p	**38** £1.30	**39** 28p	**40** 94p

REVISION

Exercise B

1 (a) 7 (b) 4 (c) 5 (d) 6

2 (a) (i) 170 (ii) 200 (b) (i) 220 (ii) 200
(c) (i) 890 (ii) 900 (d) (i) 1050 (ii) 1000

3 (a) 121 (b) 57 (c) 181 (d) 105

4 9 BCE

5 (a) $\frac{3}{4}$ (b) $\frac{2}{3}$ (c) $\frac{5}{8}$ (d) $\frac{4}{5}$

6 (a)

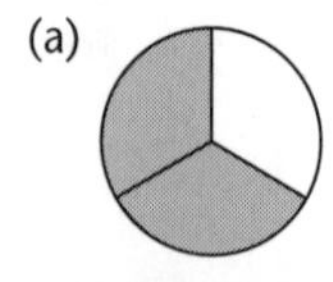

(b)

(c)

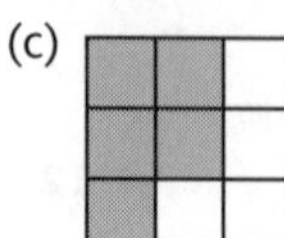

(d)

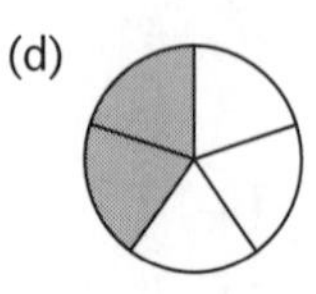

7 (a) £12 (b) 6 kg (c) 2 ml (d) £12

8 A 30%, B 25%, C 7%, D 85%

9 (a) 75% (b) 50% (c) 20% (d) 25%

10 (a) $\frac{1}{5}$ (b) $\frac{1}{10}$ (c) $\frac{1}{20}$ (d) $\frac{3}{4}$

11 (a) 8p (b) 21p (c) 20p (d) £4

Exercise BB

1 8, 2 cm
2 9 km
3 5100
4 250
5 Yes (50p change)
6 Colder, 2°
7 55p
8 85p
9 50p, £1.50
10 32p
11 232
12 $\frac{1}{4}$, $\frac{3}{4}$
13 $\frac{5}{8}$, $\frac{3}{8}$
14

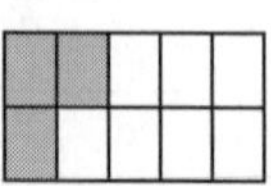

15

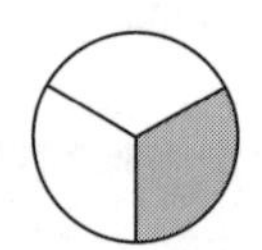

16 320 g
17 9 tonnes
18

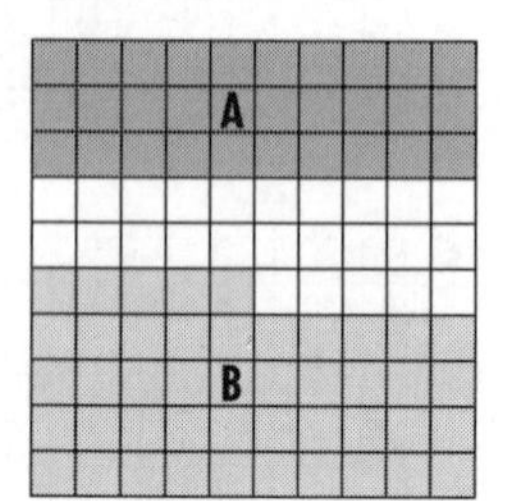

25%

19 $\frac{1}{4}$
20 20%
21 50p
22 £1.80

Algebra

31 NUMBER PATTERNS

Exercise 31A

1 10, 12
2 13, 15
3 9, 10
4 15, 18
5 45, 40
6 17, 19
7 9, 8
8 25, 30
9 35, 42
10 13, 16
11 23, 27
12 33, 35
13 30, 25
14 51, 61
15 23, 26
16 22, 20
17 5, $2\frac{1}{2}$
18 40, 42
19 16, 32
20 80, 160

Exercise 31B

1 15, 17
2 18, 21
3 54, 64
4 18, 15
5 16, 18
6 32, 37
7 28, 25
8 49, 54
9 10, 8
10 22, 15
11 12, 9
12 36, 31
13 24, 28
14 34, 40
15 5, 0
16 36, 27
17 125, 150
18 12, 17
19 4, 2
20 0, –2

Exercise 31C

1 10, 12; even numbers
2 8, 7; subtract 1
3 25, 23; subtract 2
4 32, 64; double
5 14, 16; add 2
6 10, 6; subtract 4
7 21, 24; add 3
8 27, 30; add 3
9 73, 75; add 2
10 31, 33; add 2
11 28, 34; add 6
12 11, 0; subtract 11
13 37, 46; add 9
14 25, 20; difference decreasing by 1
15 $12\frac{1}{2}$, $6\frac{1}{4}$; half
16 25, 15; subtract 10
17 58, 50; subtract 8
18 15, 21; difference increasing by 1
19 4, $4\frac{1}{2}$; add $\frac{1}{2}$
20 $\frac{1}{2}$, $\frac{1}{4}$; half

Exercise 31D

1 9, 11; add 2
2 9, 7; subtract 2
3 11, 10; subtract 1
4 23, 26; add 3
5 18, 14; subtract 4
6 54, 63; add 9
7 20, 22; add 2
8 20, 23; add 3
9 20, 24; add 4
10 24, 21; subtract 3
11 37, 45; add 8
12 28, 27; subtract 1
13 13, 15; add 2
14 9, $4\frac{1}{2}$; half
15 28, 35; difference increasing by 1
16 40, 35; difference decreasing by 1
17 20, 26; difference increasing by 1
18 25, 36; square numbers
19 65, 60; subtract 5
20 –2, –5; subtract 3

32 FUNCTION MACHINES: FINDING OUTPUTS

Exercise 32A

	(a)	(b)	(c)	(d)
1	5	9	13	17
2	5	14	23	32
3	1	5	9	13
4	8	13	18	23
5	4	10	16	22
6	7	9	11	13
7	2	7	12	17
8	26	20	14	2
9	13	21	29	37
10	0	1	2	3

Exercise 32B

	(a)	(b)	(c)	(d)
1	6	14	22	30
2	5	9	13	17
3	9	19	29	39
4	34	28	22	16
5	7	11	15	23
6	27	57	87	117

7	(a) 1	(b) 13	(c) 25	(d) 37
8	(a) 13	(b) 23	(c) 33	(d) 43
9	(a) 18	(b) 14	(c) 10	(d) 6
10	(a) 5	(b) 3	(c) 1	(d) –1

33 FUNCTION MACHINES: FINDING THE FUNCTION

Exercise 33A

1 + 1	**2** − 1	**3** + 6	**4** × 2
5 × 3	**6** − 1	**7** − 2	**8** + 1
9 + 4	**10** × 2	**11** × 4	**12** + 2
13 + 6	**14** − 3	**15** × 7	**16** × 10
17 − 3	**18** + 2	**19** + 3	**20** − 1
21 × 2	**22** + 5	**23** × 3	**24** + 3
25 ÷ 2	**26** + 10	**27** − 2	**28** − 6
29 ÷ 2	**30** × 6		

Exercise 33B

1 × 8	**2** + 4	**3** × 3	**4** + 4
5 − 4	**6** × 1	**7** × 9	**8** × 10
9 − 4	**10** + 5	**11** − 1	**12** ÷ 2
13 + 2	**14** × 4	**15** − 2	**16** + 1
17 − 1	**18** × 3	**19** × 4	**20** × 7
21 × 5	**22** + 3	**23** ÷ 5	**24** − 6
25 × 5	**26** − 8	**27** ÷ 3	**28** ÷ 2
29 + 3	**30** + 8		

34 SIMPLE EQUATIONS

Exercise 34A

1 5	**2** 2	**3** 5	**4** 3	**5** 6
6 7	**7** 4	**8** 6	**9** 5	**10** 3
11 3	**12** 10	**13** 6	**14** 4	**15** 15
16 5	**17** 4	**18** 4	**19** 8	**20** 7
21 5	**22** 6	**23** 2	**24** 3	**25** 5
26 4	**27** 8	**28** 8	**29** 9	**30** 10

Exercise 34B

1 7	**2** 4	**3** 8	**4** 19	**5** 12
6 7	**7** 12	**8** 6	**9** 4	**10** 48
11 4	**12** 11	**13** 7	**14** 24	**15** $4\frac{1}{2}$
16 9	**17** 10	**18** 8	**19** 40	**20** 5
21 18	**22** 8	**23** 42	**24** 5	**25** 9
26 5	**27** 11	**28** 4	**29** 5	**30** $3\frac{1}{2}$

35 EQUATIONS AND FORMULAE IN WORDS

Exercise 35A

1 15	**2** 60°	**3** 5
4 50 cm	**5** 4 h	**6** £8
7 8	**8** 9 cm	**9** 14°
10 20	**11** 6 years	**12** 5
13 180 min	**14** 6	**15** 8 cm
16 7	**17** 150°	**18** 16 years
19 12	**20** 60 cm	

Exercise 35B

1 12	**2** 55°	**3** 5
4 11	**5** 10	**6** £4
7 8	**8** 11	**9** 23
10 4 cm	**11** 7	**12** 4
13 8	**14** 11	**15** 6
16 15	**17** 8 years	**18** 13
19 4	**20** 3 days	

36 NAMING POINTS ON GRAPHS

Exercise 36A

1 A(1, 4), B(1, 3), C(1, 2), D(1, 1), E(2, 1), F(3, 1), G(4, 1), H(5, 1)

2 I(1, 1), J(2, 2), K(3, 3), L(4, 4), M(1, 4), N(2, 3), O(3, 2), P(4, 1)

3 Q(0, 3), R(1, 3), S(2, 3), T(3, 3), U(3, 5), V(3, 4), W(3, 2), X(3, 1)

4 A(1, 5), B(3, 5), C(3, 4), D(1, 4), E(3, 3), F(5, 3), G(5, 1), H(3, 1)

5 I(2, 0), J(3, 1), K(4, 2), L(5, 3), M(0, 5), N(1, 4), O(2, 3), P(3, 2)

6 Q(2, 2), R(3, 2), S(4, 2), T(5, 2), U(3, 5), V(4, 5), W(4, 4), X(3, 4)

Exercise 36B

1 A(1, 1), B(2, 2), C(2, 3), D(2, 4), E(3, 4), F(3, 3), G(3, 2), H(4, 1)

2 I(2, 2), J(3, 3), K(4, 4), L(4, 3), M(3, 2), N(4, 2), O(5, 3), P(3, 1)

3 Q(0, 3), R(1, 2), S(2, 3), T(3, 2), U(4, 3), V(5, 2), W(4, 1), X(2, 1)

4 A(1, 5), B(2, 5), C(2, 3), D(1, 3), E(3, 4), F(4, 4), G(4, 2), H(3, 2)

5 I(1, 5), J(2, 4), K(3, 3), L(4, 2), M(5, 1), N(0, 0), O(1, 1), P(2, 2)

6 Q(1, 4), R(3, 5), S(2, 3), T(4, 4), U(3, 2), V(5, 3), W(5, 0), X(4, 1)

37 PLOTTING POINTS ON GRAPHS

Exercise 37A

1

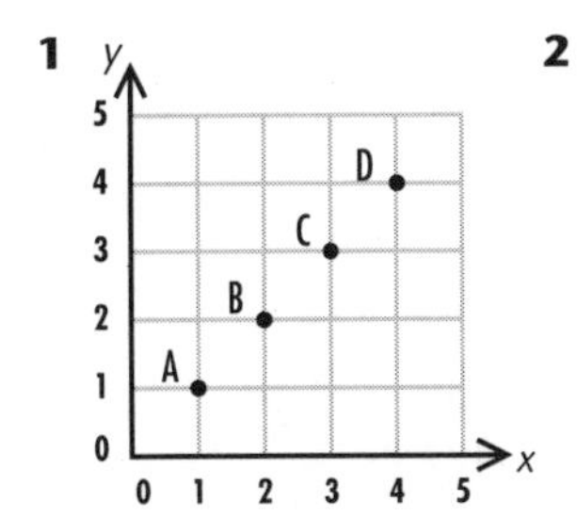

2

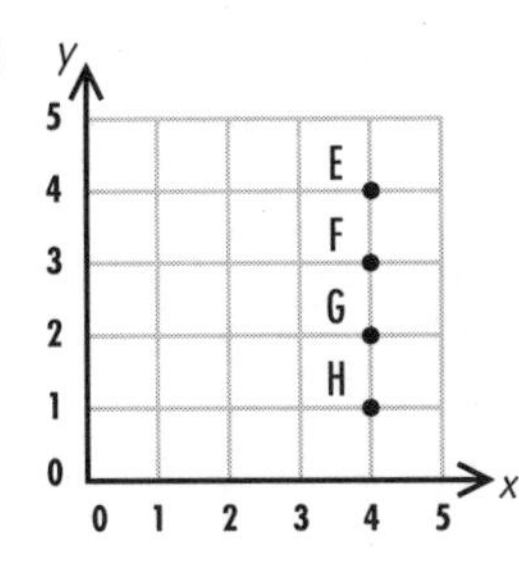

3

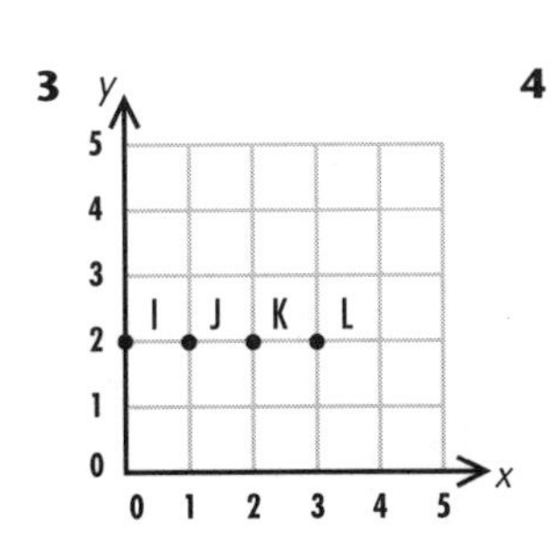

4

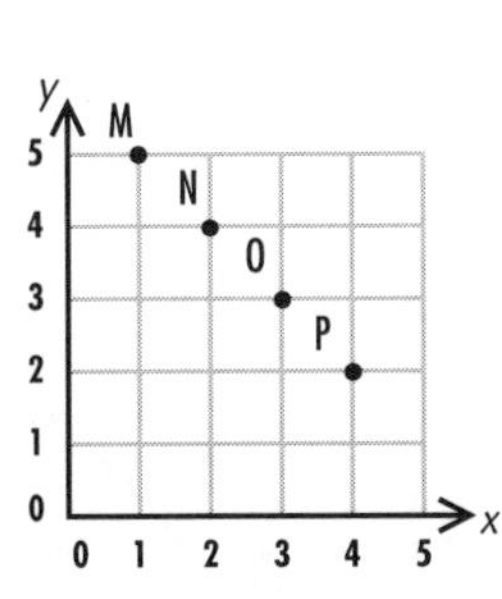

5

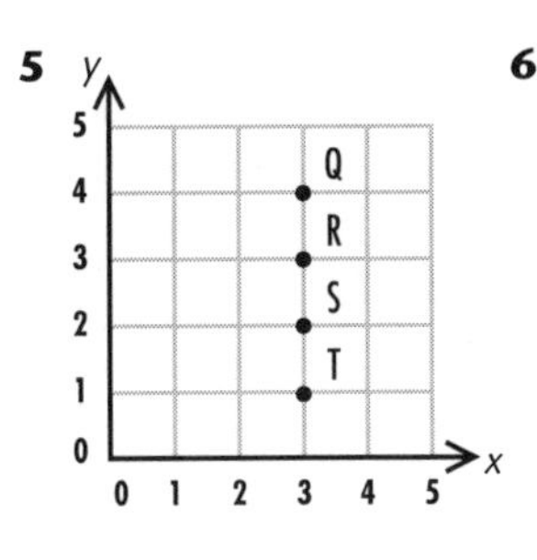

6

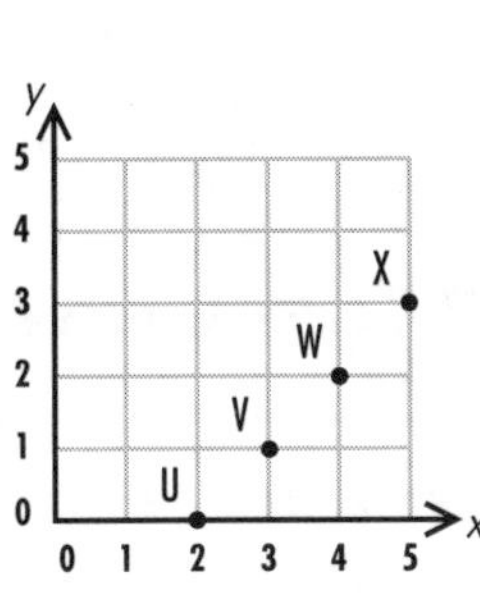

7

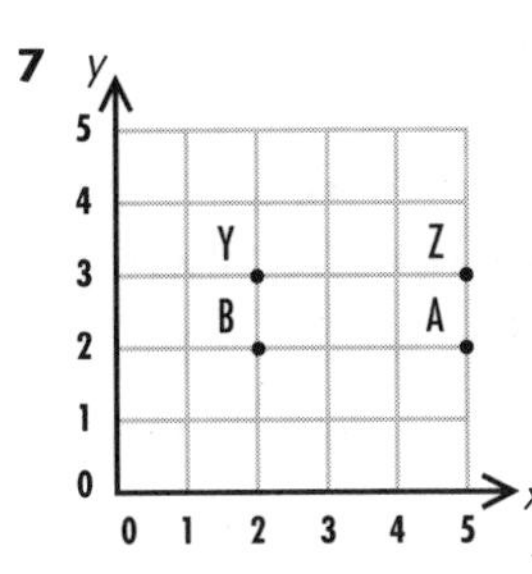

8

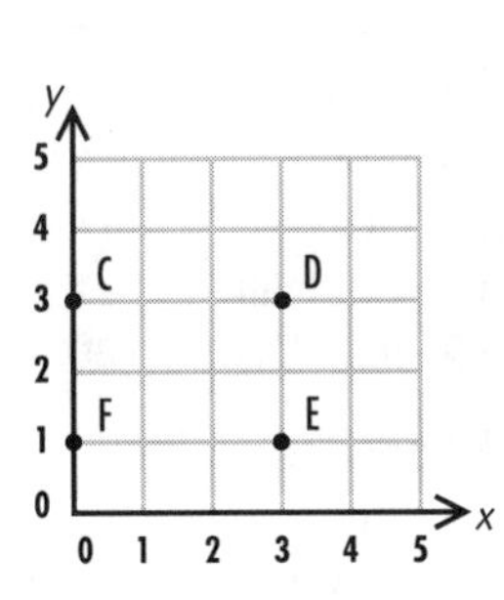

9

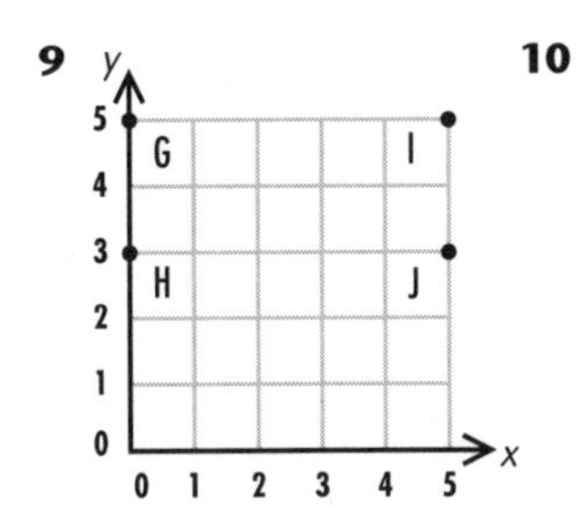

10

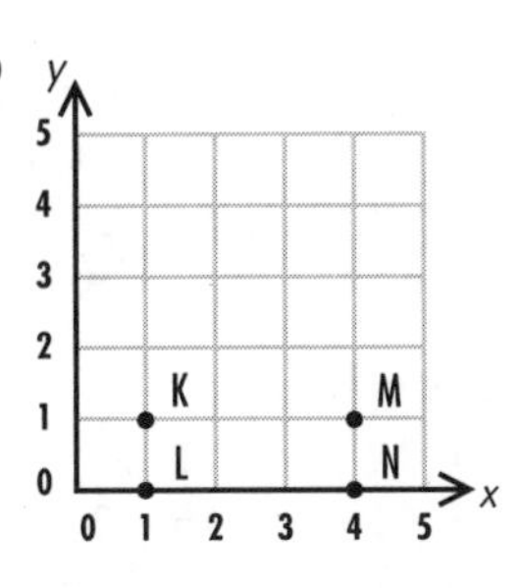

Exercise 37B

1

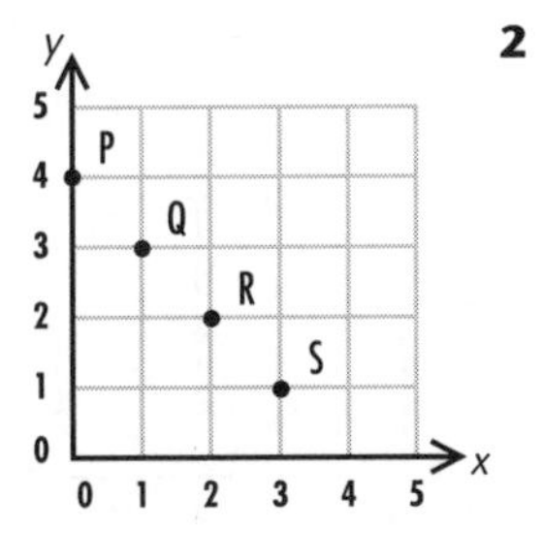

2

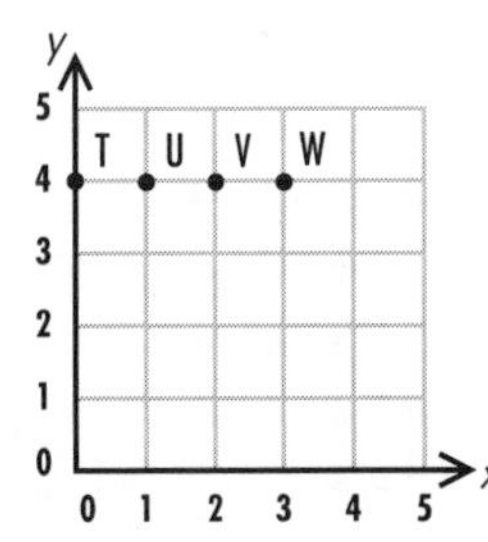

3

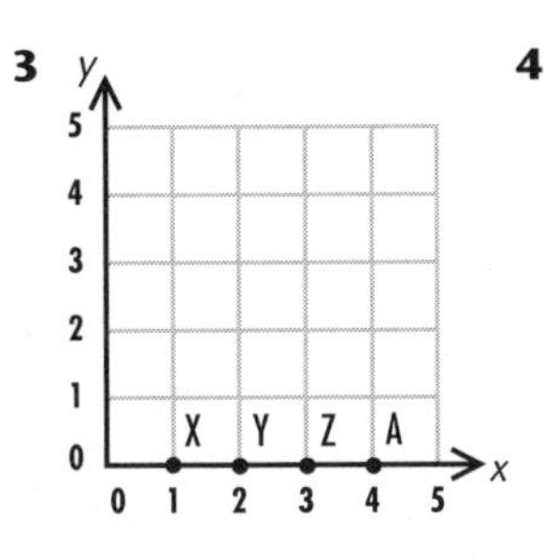

4

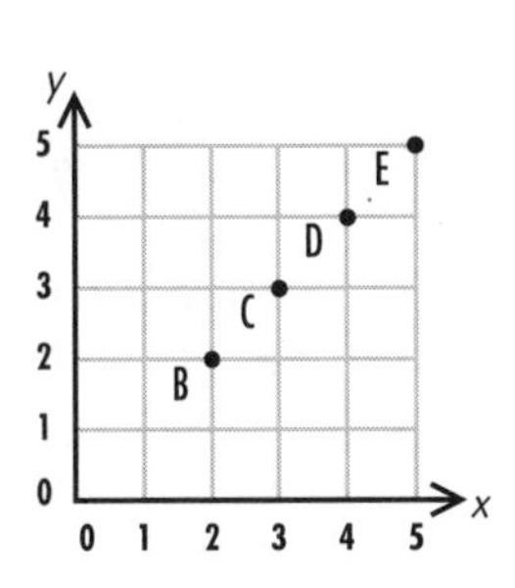

5

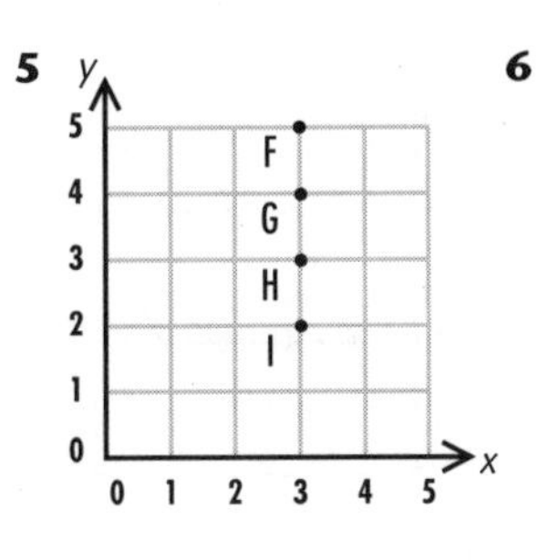

6

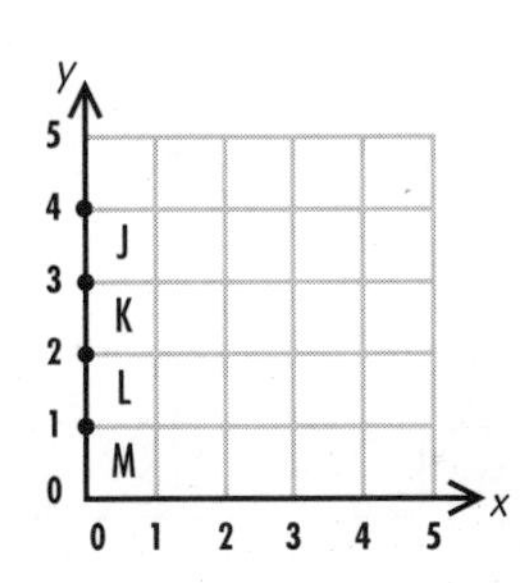

7

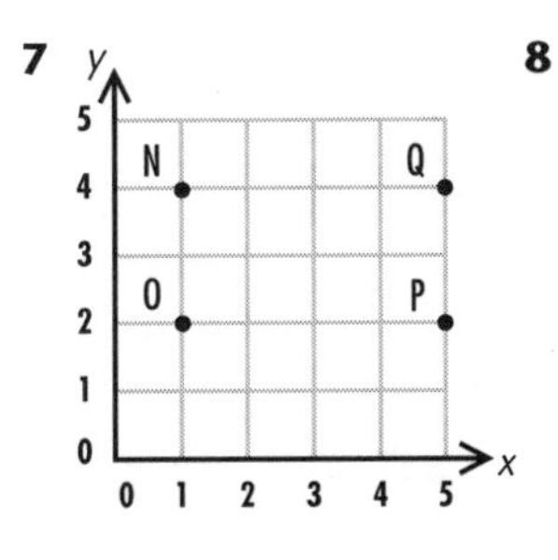

8

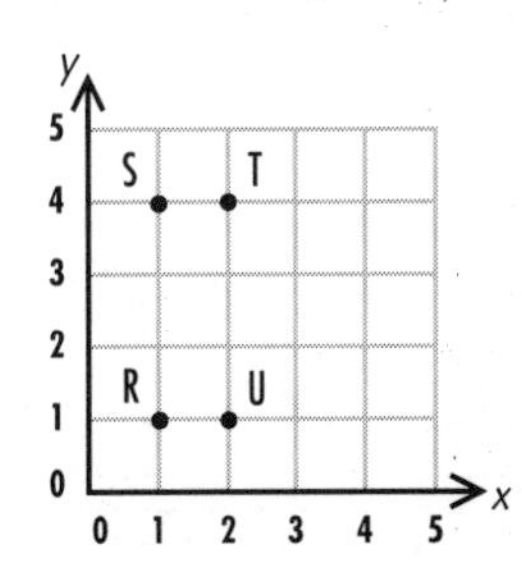

9

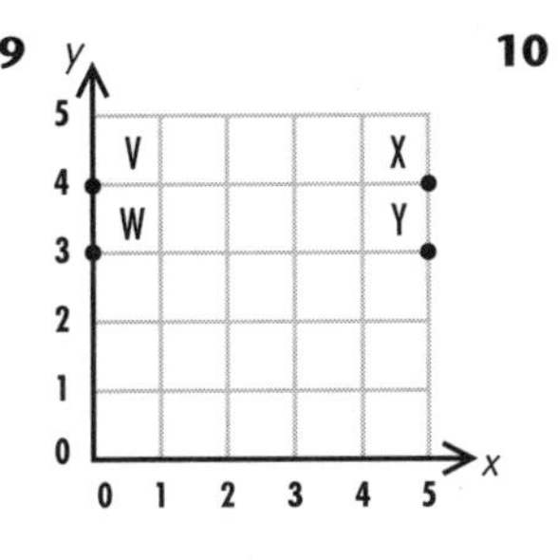

10

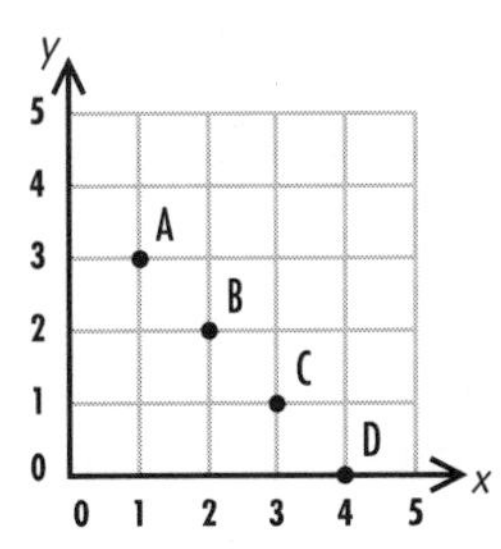

Exercise 37C

1
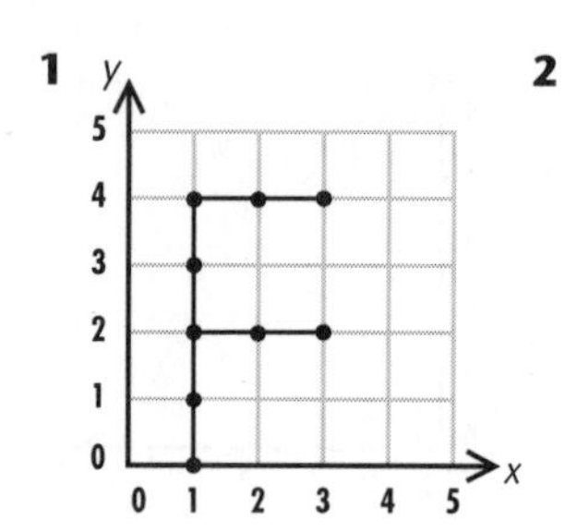

2
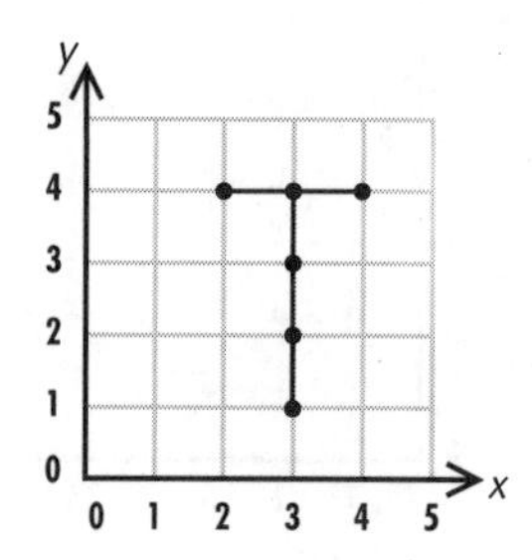

3
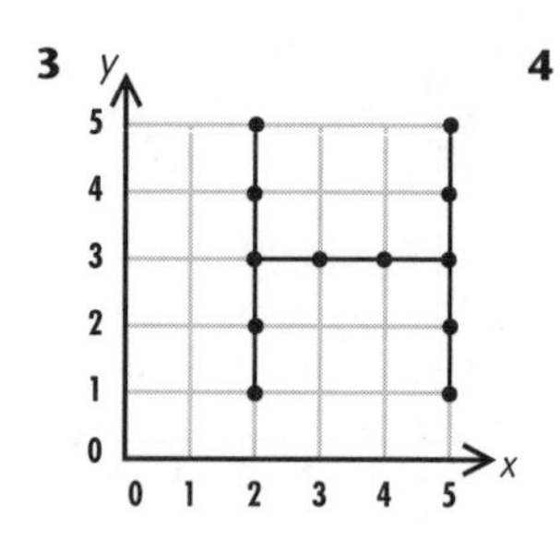

4
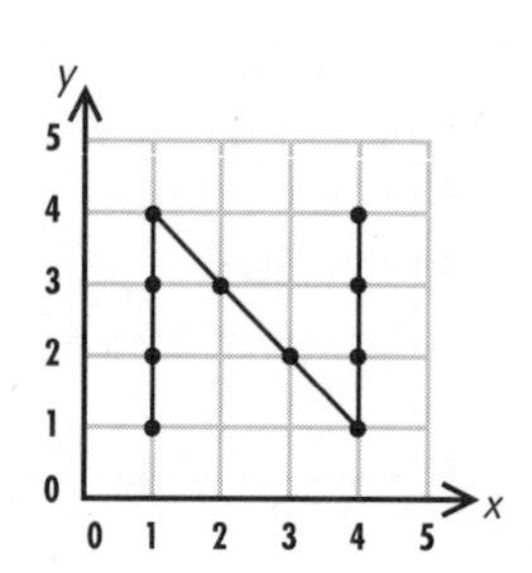

5
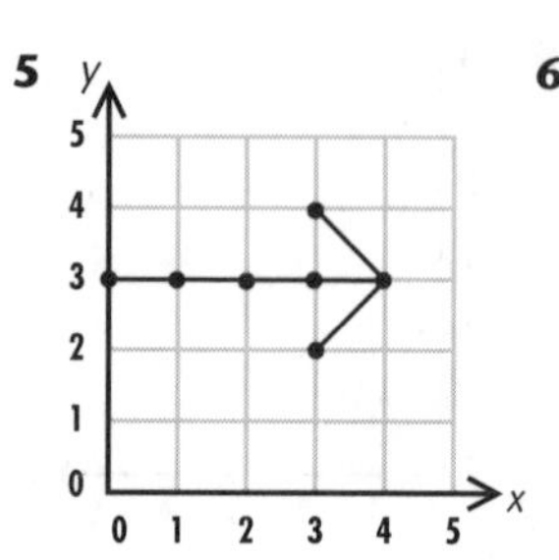

6
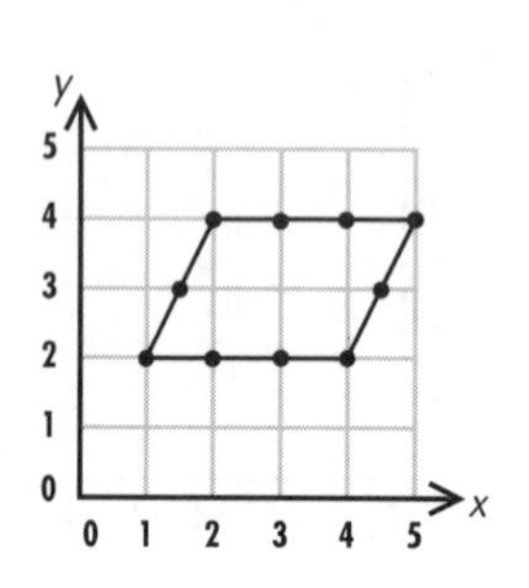

7
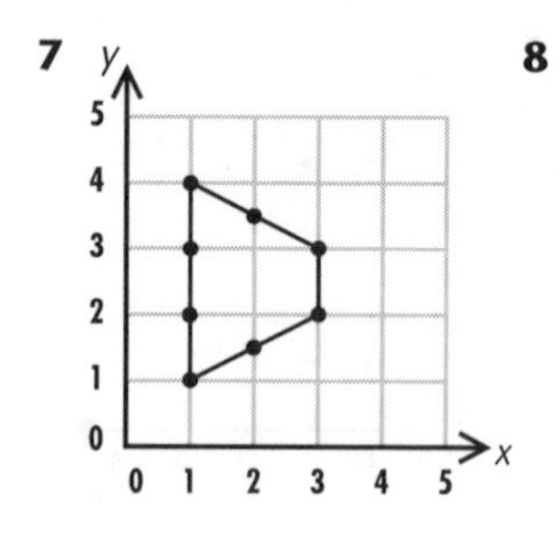

8
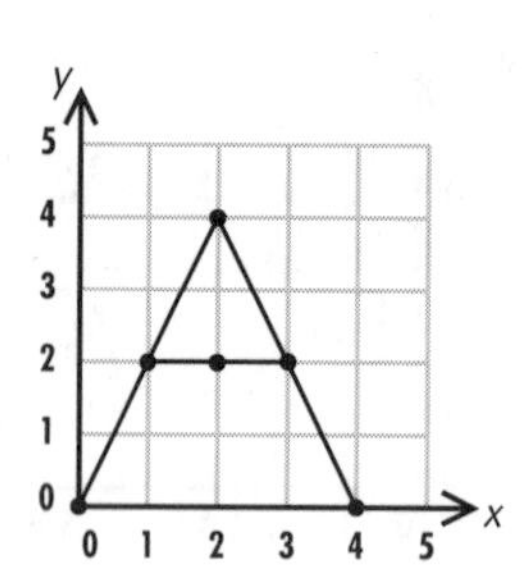

Exercise 37D

1
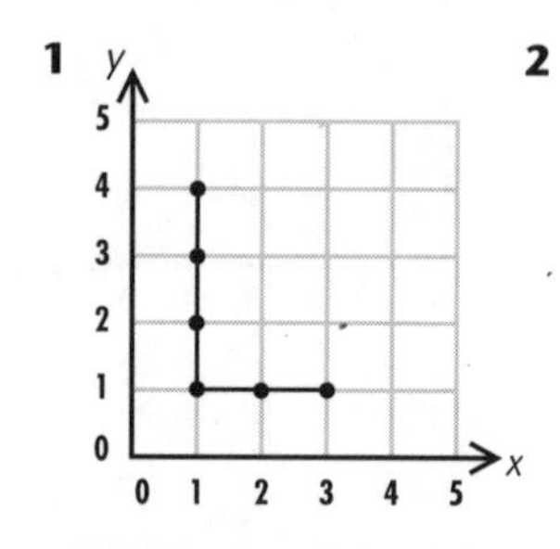

2
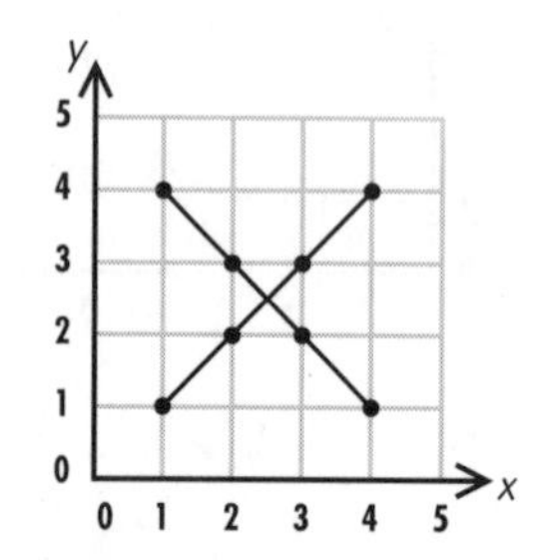

3
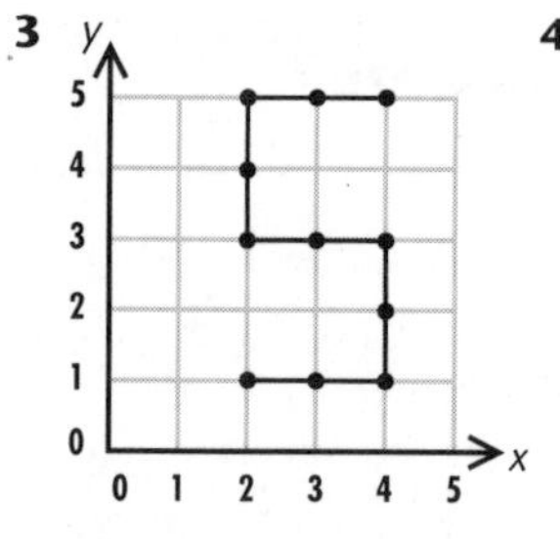

4
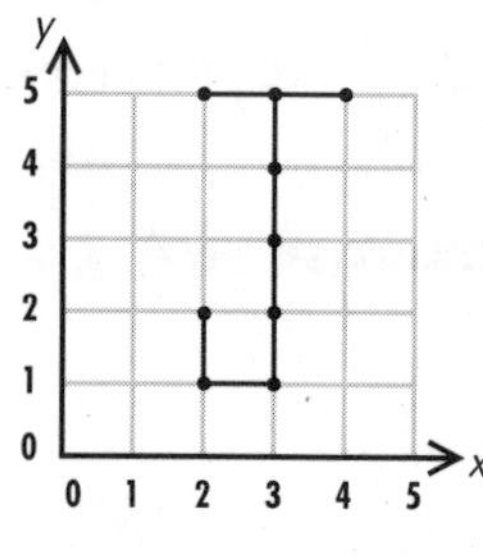

5
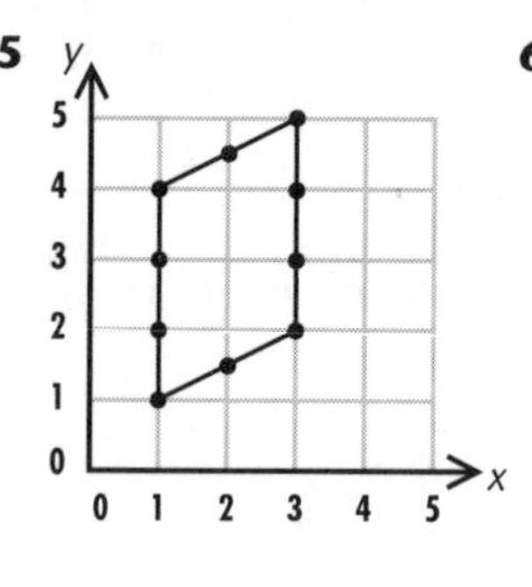

6
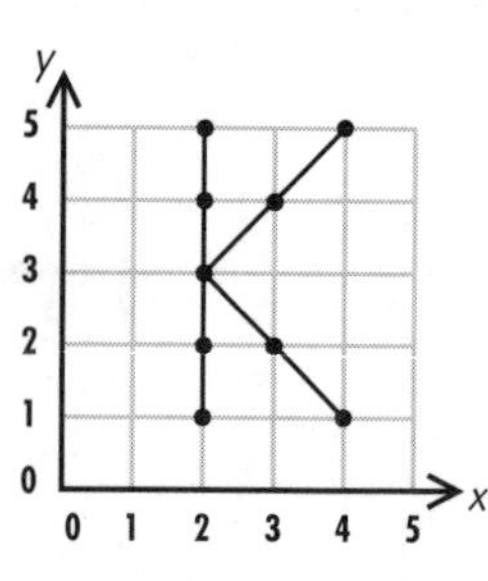

7
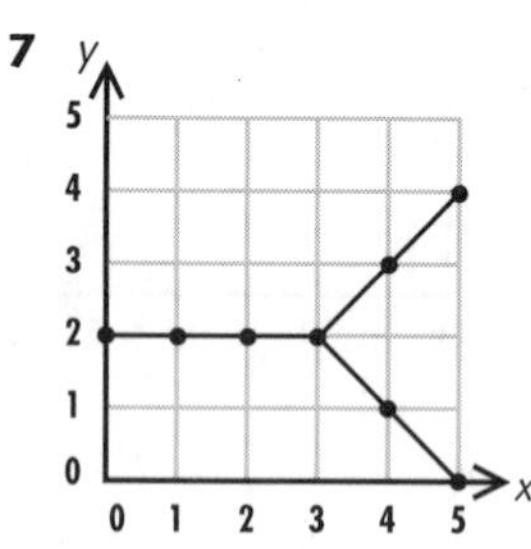

8
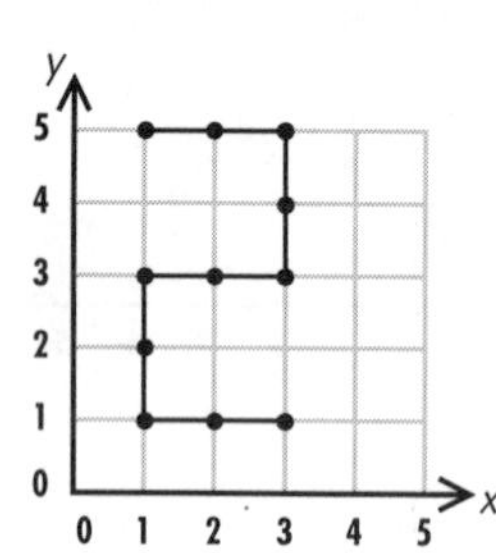

38 ROMAN NUMERALS

Exercise 38A

1 XVI **2** XXIII **3** XLV
4 XXXIV **5** XXVIII **6** LXXV
7 LXIX **8** CL **9** CC
10 CDV **11** CCCLXX **12** DCX
13 DCC **14** DCCCL **15** DCCCXXIV
16 CCXLVIII **17** DCLXXIV **18** MLXVI
19 MDCCCXII **20** MCMLXXV

Exercise 38B

1 XXV **2** XVII **3** XIX
4 XXXIII **5** XLVI **6** XXXIX
7 XXVII **8** XCII **9** CXII
10 CCC **11** CCLXXI **12** CLV
13 DCCL **14** DLIII **15** CCCXC
16 CXLIX **17** CCXCIX **18** MLXXXIX
19 MDCCCXV **20** MCMXC

Exercise 38C

1 12 **2** 8 **3** 17 **4** 41
5 51 **6** 30 **7** 29 **8** 71

9 15 **10** 26 **11** 320 **12** 510
13 28 **14** 35 **15** 60 **16** 800
17 48 **18** 2000 **19** 1801 **20** 925

Exercise 38D

1 4 **2** 11 **3** 21 **4** 14
5 7 **6** 25 **7** 9 **8** 43
9 19 **10** 84 **11** 63 **12** 230
13 140 **14** 130 **15** 555 **16** 700
17 425 **18** 900 **19** 1500 **20** 1990

REVISION

Exercise C

1 (a) 9, 11 (b) 20, 23 (c) 9, 5
(d) 27, 32

2 (a) (i) 5 (ii) 7 (iii) 9 (iv) 23
(b) (i) 2 (ii) 5 (iii) 11 (iv) 29

3 (a) – 4 (b) + 1 (c) × 2
(d) × 2 and + 1

4 (a) 5 (b) 9 (c) 6 (d) 12

5 A(1, 1), B(2, 1), C(3, 1), D(4, 2), E(2, 5), F(5, 5), G(5, 3), H(2, 3)

6 (a)

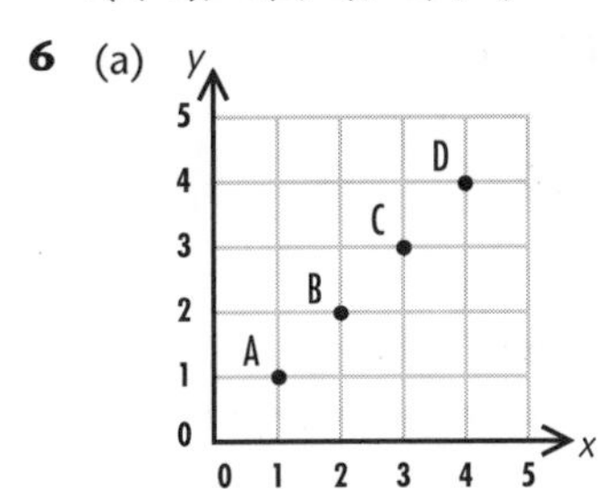

(b)

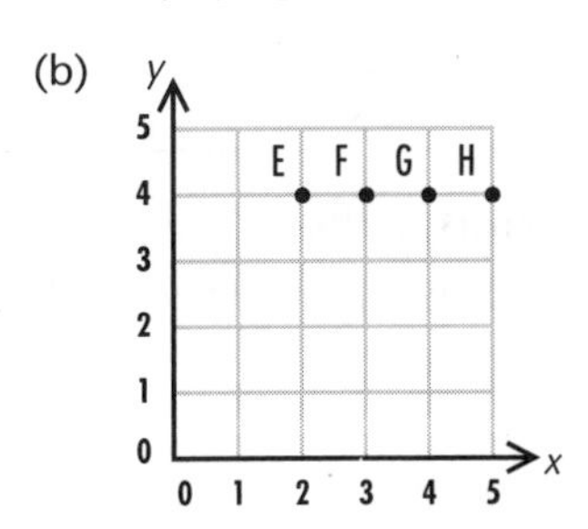

(c)

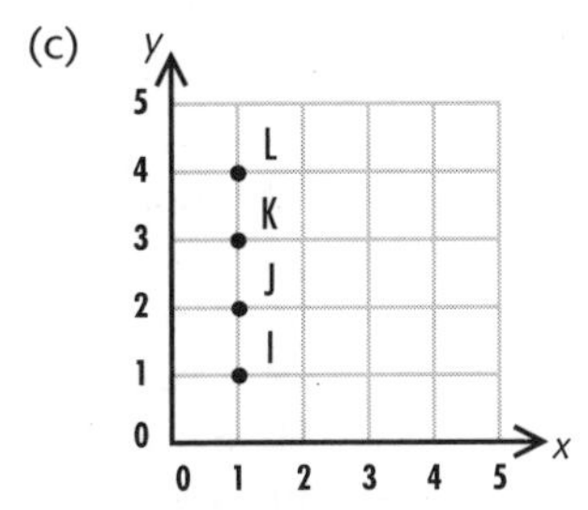

(d)

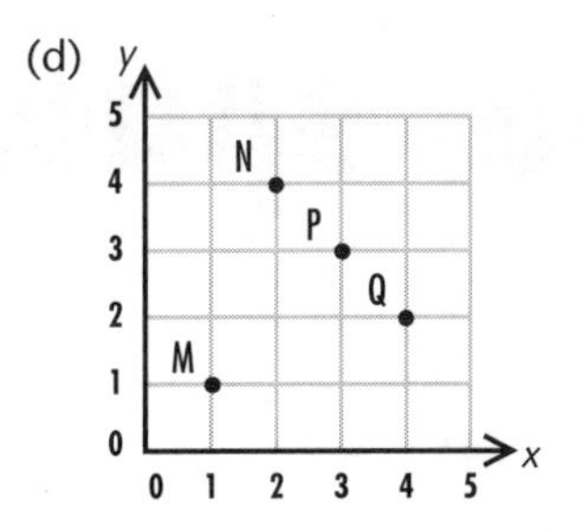

7 (a) XXXVI (b) LV (c) CCXIX
(d) DXCVIII (e) MDCCXLV

8 (a) 21 (b) 15 (c) 95 (d) 333
(e) 1928

Exercise CC

1 22, 25, 31, 37

2 9, 11

3 (a) 11, 15, 19, 25 (b) 16, 20, 24, 30; No

4 28, 7

5 (a) 3 (b) 12

6 72 cm

7 50 g

8 (a) 21 (b) 9 days

9 I(1, 1), J(1, 2), K(1, 3), L(1, 4), M(2, 4), N(3, 4), P(4, 6), Q(5, 6), R(4, 5), S(5, 5), T(3, 2), U(4, 1)

10 (a)

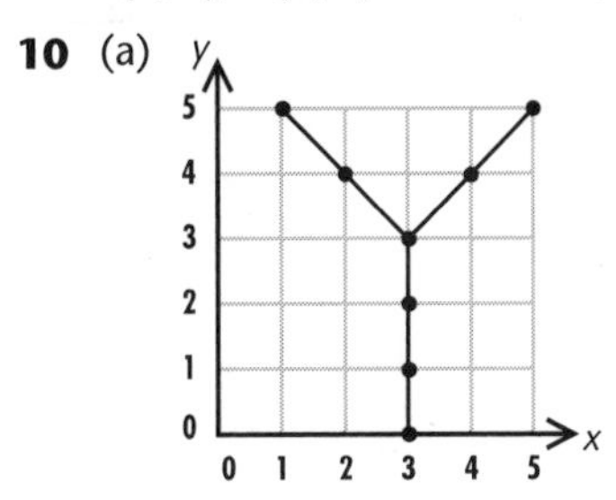

(b)

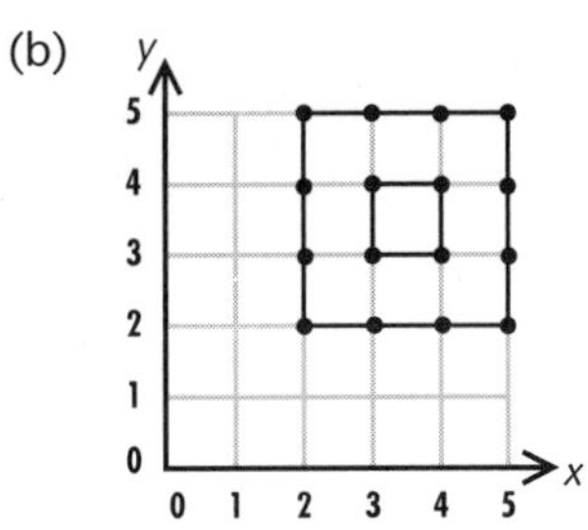

11 1918

12 MCMXCIV

Shape, space and measures

39 DIFFERENT SHAPES

Exercise 39A

1 d; rectangle
2 c; circle
3 b; square
4 a; square
5 b; not right-angled
6 c; not elliptical
7 d; 5-sided
8 c; curved edge

Exercise 39B

1 c; triangle
2 c; square
3 b; not semicircular
4 c; rectangle
5 b; square
6 a; straight edges
7 a; 5-sided
8 c; right-angled

Exercise 39C

Includes a right angle	1, 2, 4, 7, 9
3 sides	1, 8
4 sides	4, 7
Some curved sides	3, 5, 10

Exercise 39D

Includes a right angle	1, 2, 5, 7
3 sides	4, 7
4 sides	1, 3, 9
Some curved sides	2, 6, 8, 10

Exercise 39E

Some straight edges	1, 4, 5, 7, 8, 9, 10
Some curved edges	2, 3, 4, 6, 8, 10
Includes a right angle	1, 5, 7, 8

Exercise 39F

Some straight edges	1, 2, 3, 4, 5, 6, 8, 9, 10
Some curved edges	1, 7, 9
Includes a right angle	2, 3, 4, 6, 8, 10

40 MEASURING LENGTHS

Exercise 40A

1 2.5 cm
2 4.7 cm
3 2.7 cm
4 8 mm
5 3.6 cm
6 1.4 cm
7 6 mm
8 4.3 cm
9 6.9 cm
10 5.6 cm
11 2.8 cm
12 4.0 cm
13 1.1 cm
14 5.1 cm
15 1.9 cm
16 3.4 cm
17 8.2 cm
18 9.3 cm
19 7.6 cm
20 8.7 cm

Exercise 40B

1 3.8 cm
2 1.9 cm
3 7 mm
4 2.1 cm
5 3.2 cm
6 9 mm
7 4.2 cm
8 2.3 cm
9 3.9 cm
10 4.8 cm
11 1.5 cm
12 2.7 cm
13 6.0 cm
14 1.3 cm
15 5.7 cm
16 4.6 cm
17 9.4 cm
18 7.7 cm
19 8.3 cm
20 7.6 cm

41 DRAWING LINES

Exercises 41A and 41B

Check length of lines by measurement.

42 ACCURATE DRAWINGS

Exercises 42A and 42B

Check accuracy of drawings by measuring lines and angles.

43 CONGRUENT SHAPES

Exercise 43A

1 Not congruent
2 Congruent
3 Congruent
4 Not congruent
5 B and D
6 A and B
7 A and D
8 C and D
9 A and C
10 B and D
11 B and C
12 A and B

Exercise 43B

1 Not congruent
2 Congruent
3 Not congruent
4 Congruent
5 A and B
6 A and D
7 B and C
8 A and C
9 B and D
10 B and D
11 A and C
12 A and C

44 REFLECTIVE SYMMETRY

Exercise 44A

1 Y

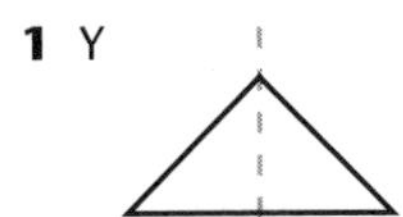

2 Y

3 N

4 Y

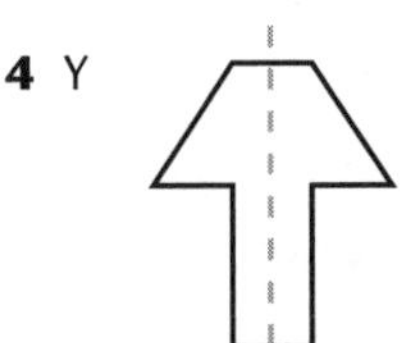

5 N

6 Y

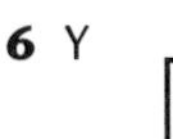

7 Y

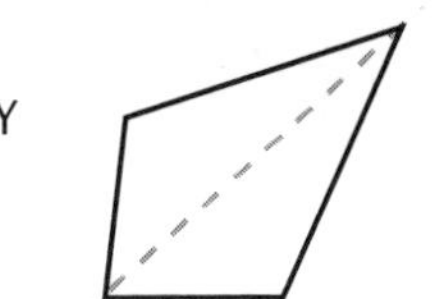

8 N

9 N

10 Y

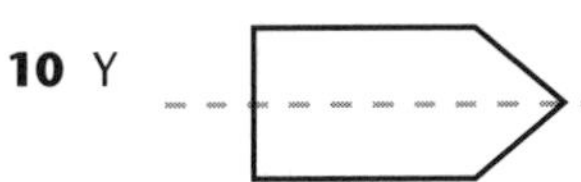

Exercise 44B

1 Y

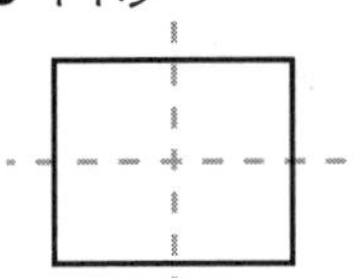

2 N

3 Y

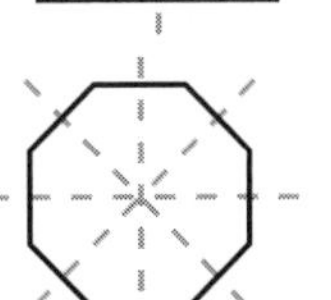

4 N

5 Y

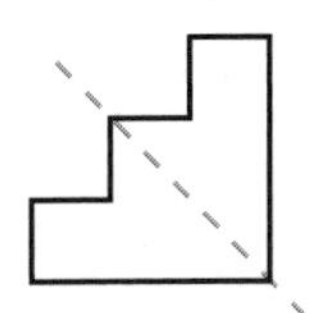

6 N

7 Y

8 Y

9 N

10 Y

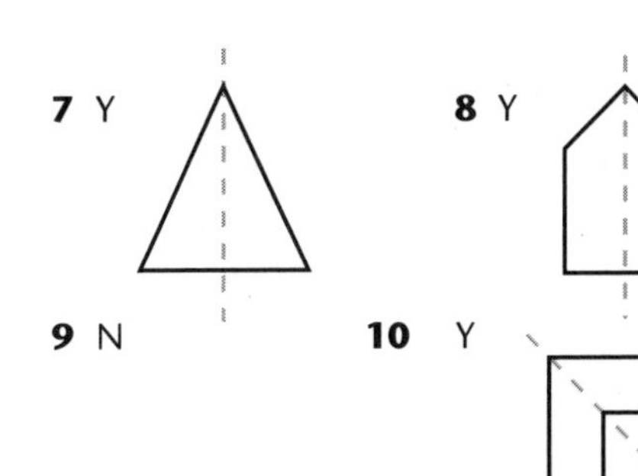

Exercise 44C

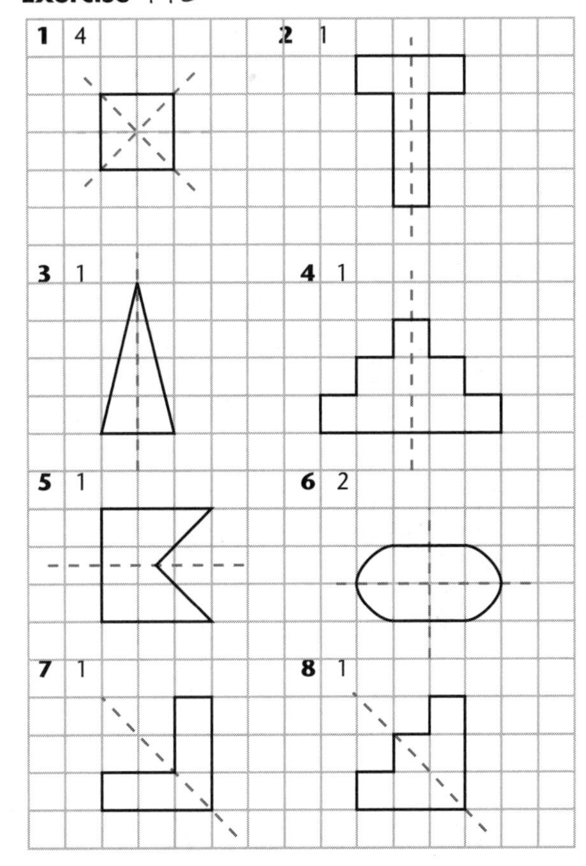

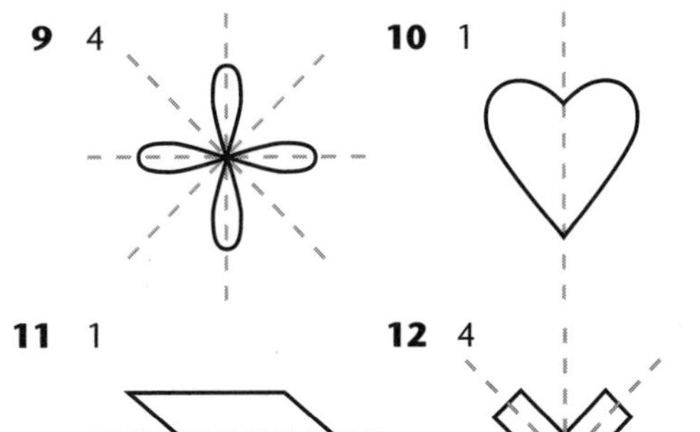

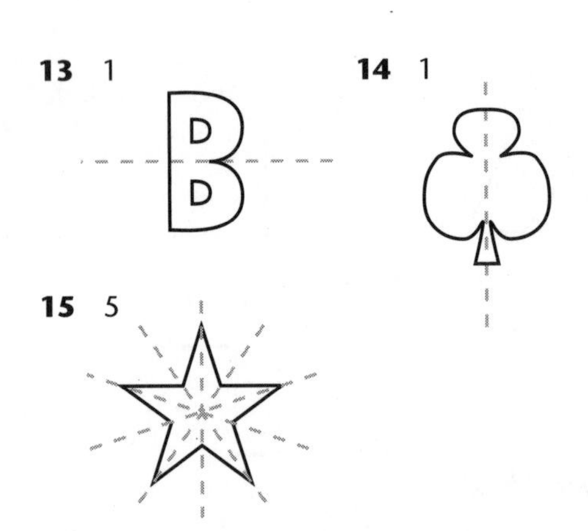

Exercise 44D

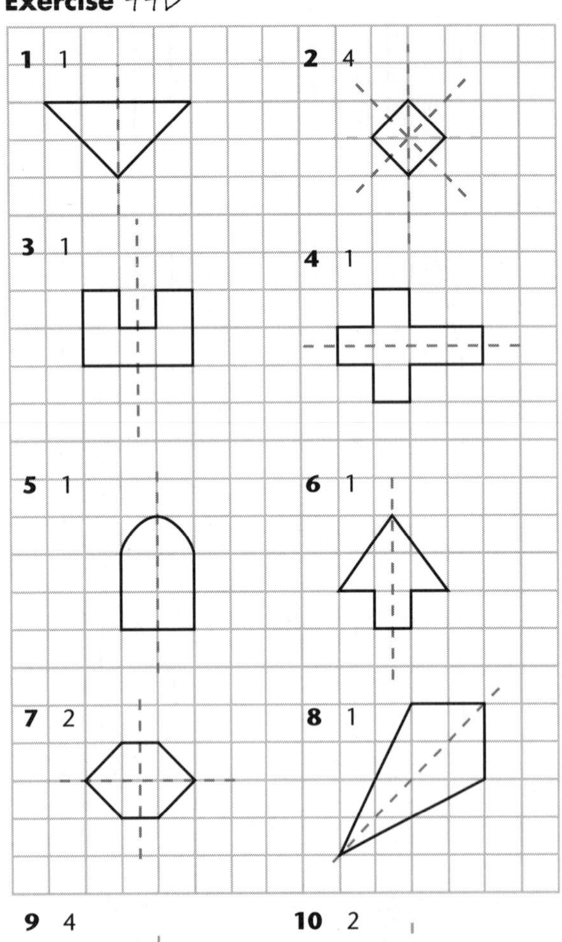

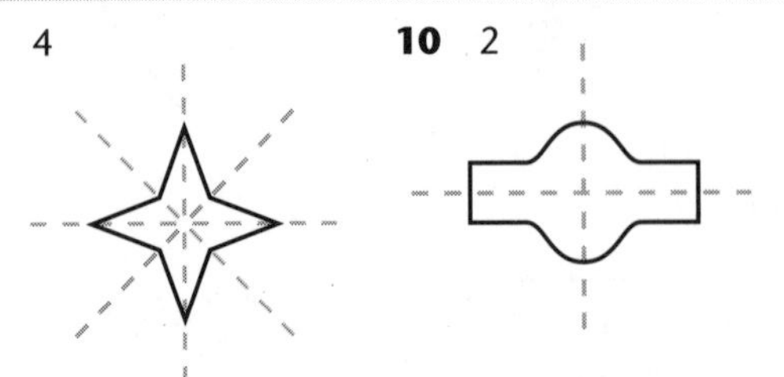

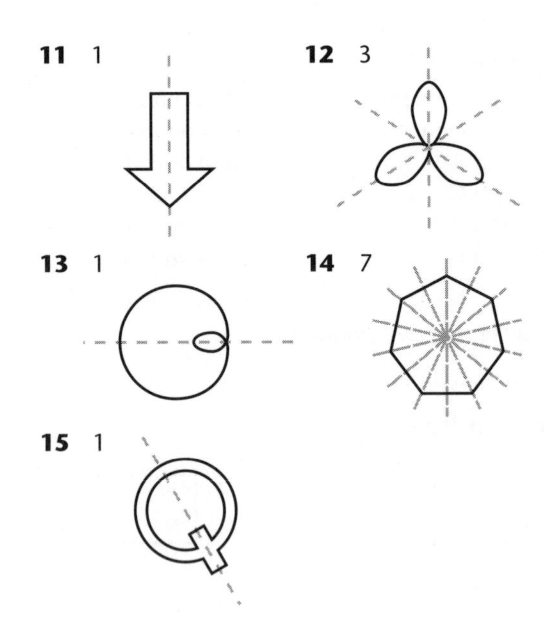

45 REFLECTION

Exercise 45A

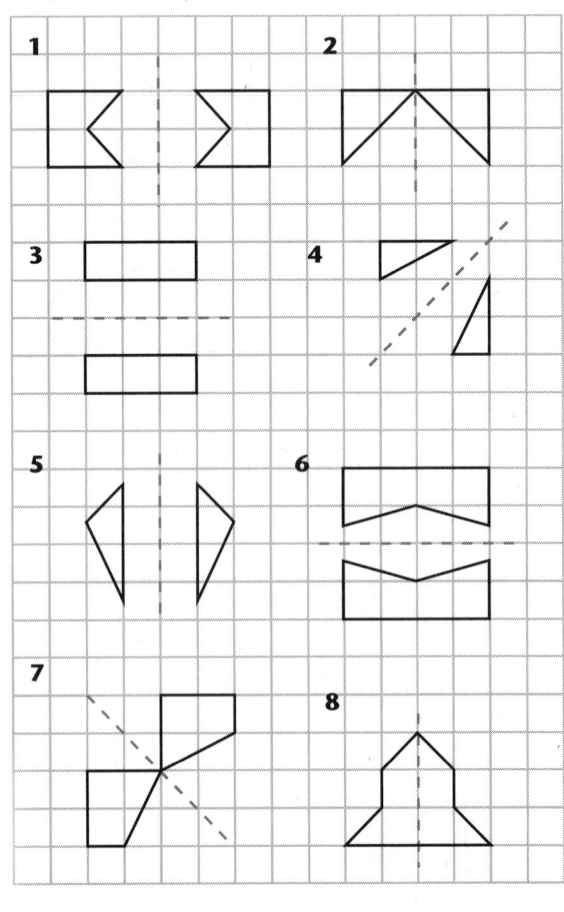

9

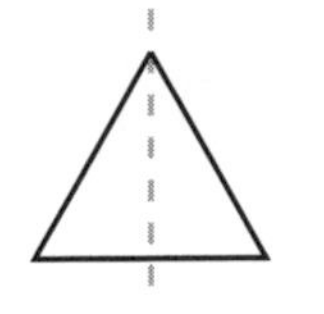

10

11

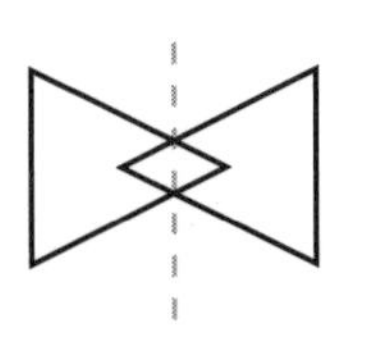

12

13

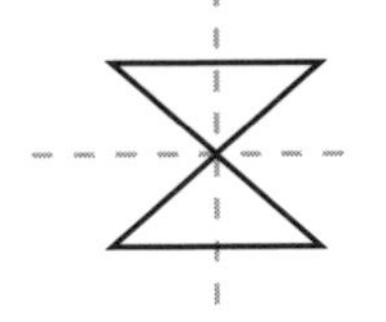

14

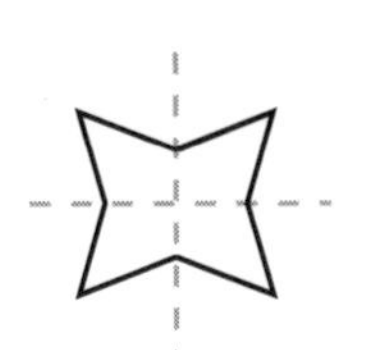

15

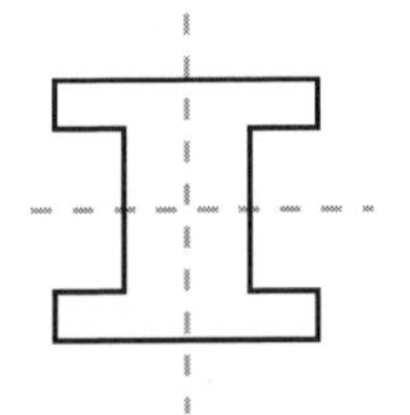

9

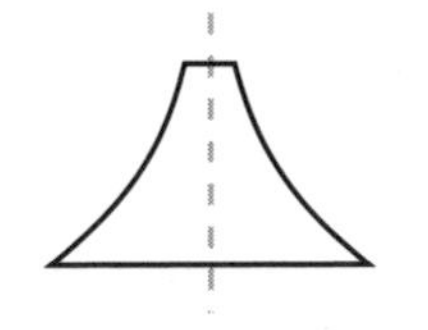

10

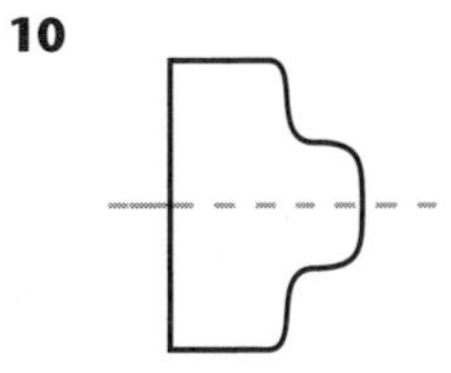

11

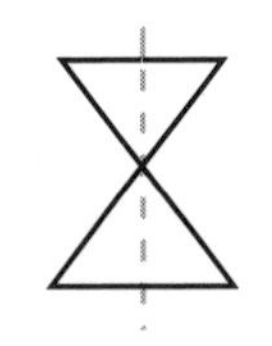

12

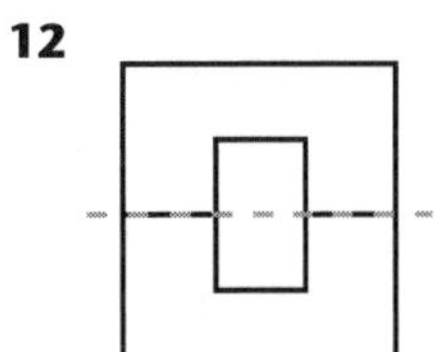

13

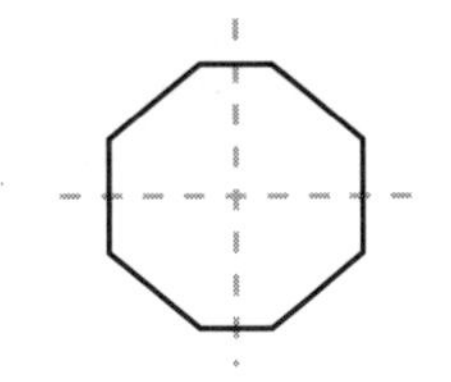

14

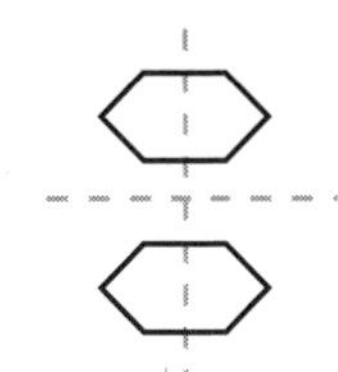

15

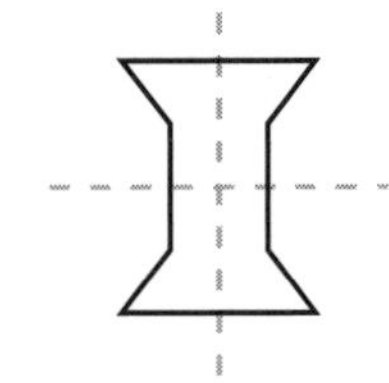

Exercise 45B

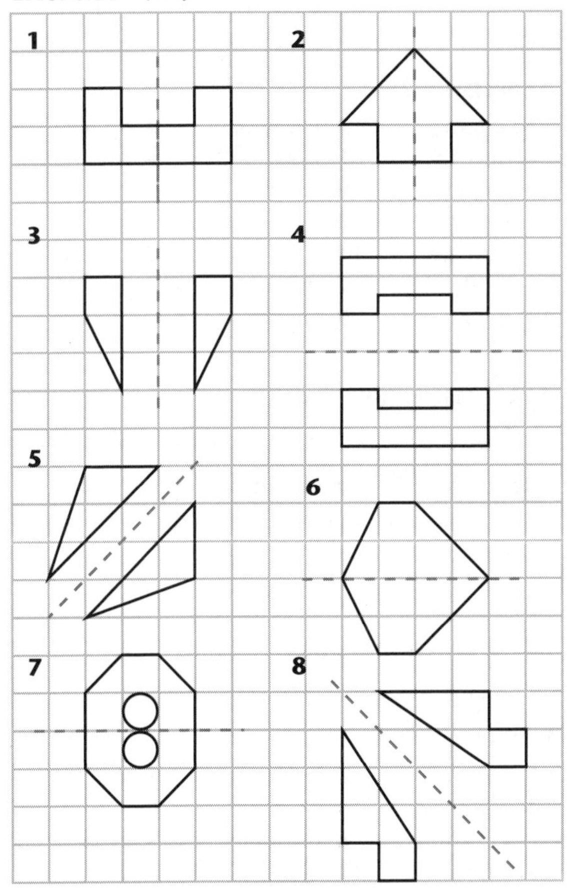

46 ROTATIONAL SYMMETRY

Exercise 46A

1 4	**2** None	**3** 8	**4** 2
5 None	**6** 2	**7** 4	**8** 3
9 None	**10** 5	**11** 8	**12** 2

Exercise 46B

1 3	**2** None	**3** 5	**4** 2
5 4	**6** None	**7** 2	**8** 4
9 6	**10** None	**11** 6	**12** None

REVISION

Exercise D

1

Includes a right angle	A, C, G, I
Some curved sides	B, C, E, F, I, J
Some straight edges	A, C, D, E, G, H, I

2 (a) 5.5 cm (b) 6.2 cm (c) 1.8 cm
(d) 3.4 cm (e) 4.4 cm (f) 3.7 cm
(g) 2.4 cm (h) 5.8 cm (i) 4.1 cm
(j) 3.2 cm

3 Check the length of the lines by measurement.

4 Check the accuracy of the drawings by measuring lines and angles.

5 (a) W, Z (b) W, X (c) X, Y (d) W, Z

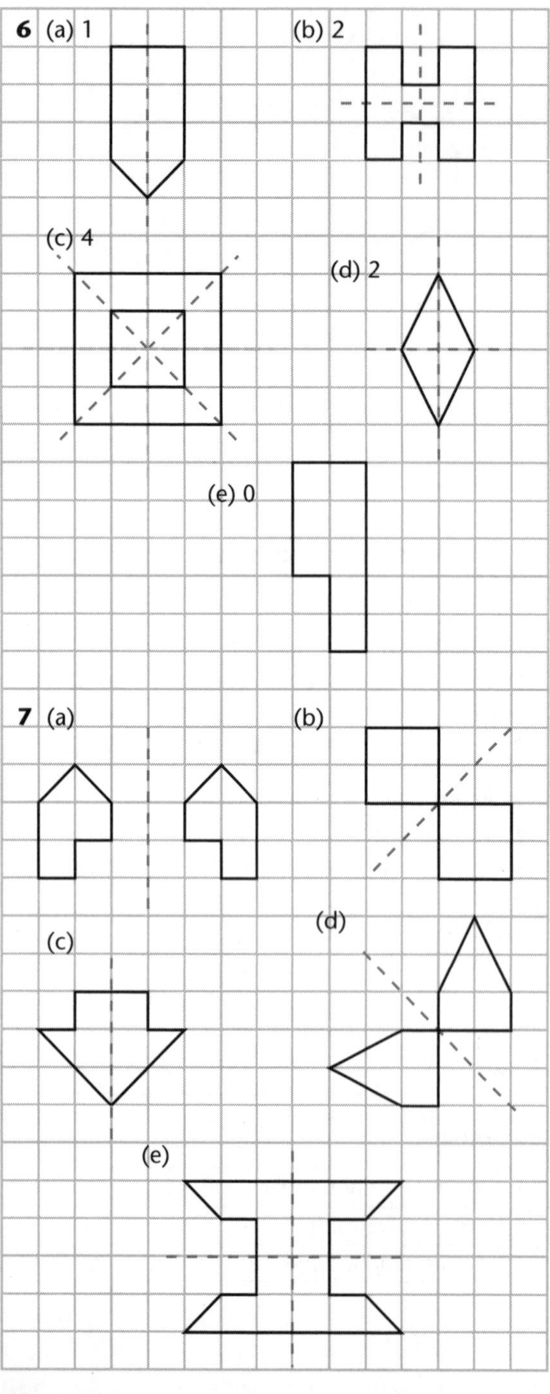

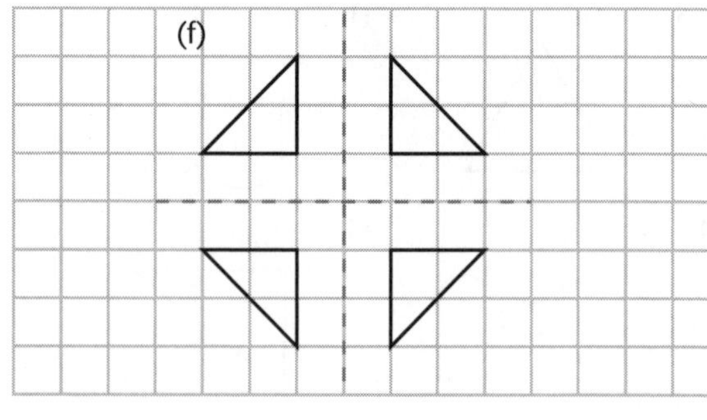

8 (a) 2 (b) 4 (c) 4 (d) 3
(e) 8 (f) 5 (g) 2 (h) 2

47 UNITS

Exercise 47A

1	mm, in	**2**	g, oz	**3**	km, m
4	cm, ft	**5**	m, ft	**6**	*l*, gal
7	kg, lb	**8**	m, ft	**9**	cl, fl. oz
10	g, oz	**11**	mm, in	**12**	cm, in
13	cl, fl. oz	**14**	kg, st	**15**	t, t
16	m, ft	**17**	mm, in	**18**	mg, oz
19	cl, pt	**20**	ml, fl. oz		

Exercise 47B

1	km, m	**2**	ml, fl. oz	**3**	kg, lb
4	mm, in	**5**	*l*, gal	**6**	mg, oz
7	cm, in	**8**	kg, lb	**9**	*l*, gal
10	cm, ft	**11**	m, ft	**12**	km, m
13	cl, fl. oz	**14**	t, t	**15**	cm, in
16	m, ft	**17**	mg, oz	**18**	*l*, pt
19	g, oz	**20**	km, m		

48 ESTIMATION

Exercises 48A and 48B

Check all estimates by actual measurement where possible.

49 SCALES AND DIALS

Exercise 49A

1	109 g	**2**	1.5 kg	**3**	2.6
4	3.5	**5**	16.2	**6**	4.8
7	2.2 amps	**8**	0.2 amps	**9**	27 g
10	6.8 kg	**11**	170		

12 (a) 140 km/h (b) 90 m.p.h.
13 (a) 100 km/h (b) 65 m.p.h.
14 (a) 156 km/h (b) 101 m.p.h.
15 (a) 86 km/h (b) 55 m.p.h.

Exercise 49B

1 160 g **2** 26 g **3** 6.6
4 3.4 **5** 7.8 **6** 4.4
7 1.7 amps **8** 0.8 amps **9** 23 g
10 35 g **11** 2.5 *l*
12 (a) 18°C (b) 64°F **13** (a) 32°C (b) 89°F
14 (a) 31°C (b) 87°F **15** (a) 27°C (b) 80°F

50 TIME

Exercise 50A

1 9.00; nine o'clock
2 7.00; seven o'clock
3 8.30; half past eight
4 1.15; quarter past one
5 11.45; quarter to twelve
6 3.30; half past three
7 6.15; quarter past six
8 2.40; twenty to three
9 3.45; quarter to four
10 7.10; ten past seven

Exercise 50B

1 5.00; five o'clock
2 2.00; two o'clock
3 8.15; quarter past eight
4 3.30; half past three
5 5.15; quarter past five
6 10.45; quarter to eleven
7 2.30; half past two
8 1.25; twenty-five past one
9 7.15; quarter past seven
10 7.50; ten to eight

Exercise 50C

1 0900 h **2** 1030 h **3** 1115 h
4 1540 h **5** 1820 h **6** 0810 h
7 1240 h **8** 2045 h **9** 1725 h
10 1515 h **11** 1430 h **12** 0725 h
13 1130 h **14** 1155 h **15** 2115 h
16 12.55 p.m. **17** 11.10 a.m. **18** 8.10 a.m.
19 00.25 a.m. **20** 7.40 a.m. **21** 11.35 p.m.
22 2.15 p.m. **23** 10.35 a.m. **24** 8.55 a.m.
25 9.20 a.m. **26** 10.25 p.m. **27** 8.55 p.m.
28 6.16 p.m. **29** 5.25 p.m. **30** 1.20 p.m.

Exercise 50D

1 0950 h **2** 1015 h **3** 1215 h
4 1425 h **5** 2210 h **6** 0720 h
7 0935 h **8** 1315 h **9** 0830 h
10 2325 h **11** 1450 h **12** 1250 h
13 0150 h **14** 0020 h **15** 1220 h
16 2.55 a.m. **17** 12.10 p.m. **18** 00.10 a.m.
19 2.30 p.m. **20** 11.50 p.m. **21** 0.54 a.m.
22 7.30 a.m. **23** 3.30 a.m. **24** 1.10 p.m.
25 8.20 a.m. **26** 1.45 p.m. **27** 3.42 a.m.
28 10.14 p.m. **29** 9.17 a.m. **30** 1.27 p.m.

Exercise 50E

1 2130 h **2** 11.00 a.m.
3 1115 h **4** 9.05 p.m.
5 6.25 a.m. **6** 1700 h
7 0300 h **8** 10.00 a.m.
9 3.35 p.m. **10** 1110 h
11 1955 h **12** 10.10 a.m.
13 1 h 25 min **14** 50 min
15 $11\frac{1}{2}$ h **16** 14 h 20 min
17 5 h 10 min **18** 3 h 15 min
19 19 h 32 min **20** 1 h 53 min
21 4 h 37 min **22** 21 h 30 min
23 3 h 25 min **24** 8 h 56 min
25 4 h 20 min

Exercise 50F

1 4.15 a.m. **2** 0125 h
3 6.35 p.m. **4** 1245 h
5 0640 h **6** 1940 h
7 1150 h **8** 7.30 a.m.
9 6.55 p.m. **10** 1805 h
11 0455 h **12** 3.50 p.m.
13 4 h 10 min **14** 2 h 15 min
15 3 h 55 min **16** 4 h 40 min
17 6 h 45 min **18** 2 h 45 min
19 6 h 6 min **20** 3 h 10 min
21 5 h 20 min **22** 16 h 20 min
23 9 h 15 min **24** 5 h 10 min
25 4 h 5 min

51 AREA AND PERIMETER

Exercise 51A

1 (a) 8 cm (b) 4 cm^2
2 (a) 10 cm (b) 6 cm^2
3 (a) 6 cm (b) 2 cm^2
4 (a) 12 cm (b) 8 cm^2

5 (a) 8 cm (b) 3 cm^2
6 (a) 10 cm (b) 4 cm^2
7 (a) 12 cm (b) 8 cm^2
8 (a) 12 cm (b) 5 cm^2
9 (a) 12 cm (b) 7 cm^2
10 (a) 18 cm (b) 8 cm^2
11 (a) 16 cm (b) 13 cm^2
12 (a) 16 cm (b) 11 cm^2
13 (a) 14 cm (b) 10 cm^2
14 (a) 14 cm (b) 12 cm^2
15 (a) 18 cm (b) 20 cm^2
16 (a) 16 cm (b) 16 cm^2
17 (a) 18 cm (b) 15 cm^2
18 (a) 20 cm (b) 16 cm^2
19 (a) 20 cm (b) 16 cm^2
20 (a) 18 cm (b) 14 cm^2

Exercise 51B

1 (a) 10 cm (b) 4 cm^2
2 (a) 12 cm (b) 8 cm^2
3 (a) 12 cm (b) 9 cm^2
4 (a) 8 cm (b) 3 cm^2
5 (a) 8 cm (b) 4 cm^2
6 (a) 10 cm (b) 5 cm^2
7 (a) 12 cm (b) 6 cm^2
8 (a) 12 cm (b) 6 cm^2
9 (a) 14 cm (b) 8 cm^2
10 (a) 14 cm (b) 7 cm^2
11 (a) 14 cm (b) 10 cm^2
12 (a) 16 cm (b) 8 cm^2
13 (a) 16 cm (b) 15 cm^2
14 (a) 22 cm (b) 30 cm^2
15 (a) 20 cm (b) 25 cm^2
16 (a) 20 cm (b) 24 cm^2
17 (a) 18 cm (b) 16 cm^2
18 (a) 24 cm (b) 29 cm^2
19 (a) 28 cm (b) 24 cm^2
20 (a) 18 cm (b) 14 cm^2

52 VOLUME

Exercise 52A

1 15 cm^3 2 12 cm^3 3 10 cm^3
4 4 cm^3 5 13 cm^3 6 10 cm^3
7 11 cm^3 8 12 cm^3 9 14 cm^3
10 18 cm^3 11 32 cm^3 12 41 cm^3
13 36 cm^3 14 60 cm^3 15 30 cm^3

Exercise 52B

1 12 cm^3 2 9 cm^3 3 6 cm^3
4 12 cm^3 5 12 cm^3 6 8 cm^3
7 12 cm^3 8 13 cm^3 9 15 cm^3
10 27 cm^3 11 30 cm^3 12 60 cm^3
13 34 cm^3 14 57 cm^3 15 36 cm^3

REVISION

Exercise E

1 (a) kg (b) ml (c) cm (d) m
(e) mm (f) t (g) *l* (h) g
2 (a) 12.5°C (b) 9 amps (c) 170 ml (d) 240 g
(e) 1.25 (f) 2.8 (g) 32 amps
3 (a) (i) 0500 hours (ii) Five o'clock
(b) (i) 0715 hours (ii) Quarter past seven
(c) (i) 0120 hours (ii) Twenty past one
(d) (i) 1145 hours (ii) Quarter to twelve
(e) (i) 0650 hours (ii) Ten to seven
4 (a) 0740 h (b) 1730 h
(c) 0425 h (d) 2055 h
(e) 2305 h
5 (a) 1.40 a.m. (b) 1.30 p.m.
(c) 10.10 a.m. (d) 10.05 p.m.
(e) 3.15 p.m.
6 (a) 2 h 5 min (b) 4 h 20 min
(c) 9 h 20 min (d) 3 h 25 min
(e) 9 h 39 min
7 (a) (i) 12 cm (ii) 9 cm^2
(b) (i) 12 cm (ii) 8 cm^2
(c) (i) 12 cm (ii) 7 cm^2
(d) (i) 10 cm (ii) 4 cm^2
(e) (i) 18 cm (ii) 8 cm^2
8 (a) 36 cm^3 (b) 16 cm^3 (c) 30 cm^3
(d) 36 cm^3 (e) 72 cm^3

Exercise EE

1 (a) 17 (b) 45 min (c) 40 min
(d) 0833 (e) 1725 or 5.25 p.m.
2 (a) 9 (b) 40 m^2 (c) 26
(d) 14 (e) 28 m
3 (a) 36 m^2 (b) 30 m (c) £105

Handling data

53 USING TABLES

Exercise 53A

1 £309 **2** £239 **3** £529
4 £429 **5** £359 **6** £509
7 £429 **8** £429 **9** £14 995
10 £12 295 **11** £17 495 **12** £10 995
13 £13 995 **14** £10 995 **15** £30 995
16 £16 995 **17** 730/1427 **18** 730/1582
19 730/1513 **20** 730/1489 **21** 730/1427
22 730/1489 **23** 730/1544 **24** 730/1702
25 730/1458

Exercise 53B

1 £339 **2** £339 **3** £319
4 £429 **5** £339 **6** £319
7 £389 **8** £339 **9** £6909.84
10 £1089.96 **11** £13.95 **12** £3445.68
13 £239.17 **14** £19.35 **15** £2180.04
16 £262.71 **17** £14.75 and £17.75
18 £7.75 **19** £14.75 **20** £89.00
21 £17.75 **22** £12.75
23 £6.75 and £10.75 **24** £14.75
25 £49.50

54 USING TIMETABLES

Exercise 54A

1 1102 **2** 2114 **3** 2254
4 14 min **5** 49 min **6** 17 min
7 2048 **8** 1414 **9** 1143
10 0937 **11** 0840 **12** 1045
13 1247 **14** 1 h 22 min **15** 2 h 38 min
16 1 h 22 min **17** 0842 **18** 0727
19 0730 **20** 1350

Exercise 54B

1 0800 **2** 2205 **3** 0838
4 15 min **5** 33 min **6** 36 min
7 0807 **8** 0810 **9** 2148
10 1152 **11** 1930 **12** 1930
13 2055 **14** 1 h 8 min **15** 43 min
16 1 h 30 min **17** 1827 **18** 1838
19 1800 **20** 1912

55 MAKING FREQUENCY TABLES

Exercise 55A

1

Matches	Frequency
38	6
39	8
40	9
41	7
42	5
Total	35

2

Vowel	Frequency
a	9
e	11
i	7
o	8
u	5
Total	40

3

Numbers	Frequency
1	5
2	6
3	8
4	5
5	7
6	5
Total	36

4

Brothers/sisters	Frequency
0	4
1	5
2	8
3	6
4	4
5	3
Total	30

5

Letter	Frequency
R	4
S	6
T	6
U	8
V	9
W	7
Total	40

6

Breakdowns	Frequency
0	5
1	8
2	10
3	6
4	5
5	2
Total	36

7

Defects	Frequency
2	2
3	5
4	7
5	8
6	6
7	4
Total	32

8

Guinea pigs	Frequency
1	4
2	3
3	8
4	7
5	9
6	4
Total	35

9

Deliveries	Frequency
4	4
5	7
6	12
7	9
8	7
9	6
Total	45

10

Ages	Frequency
11	7
12	9
13	6
14	7
15	6
16	5
Total	40

Exercise 55B

1

Sweets	Frequency
23	5
24	7
25	9
26	6
27	3
Total	30

2

Shoe size	Frequency
3	3
4	8
5	6
6	8
7	3
8	2
Total	30

3

Pupil late	Frequency
A	6
B	8
C	3
D	8
E	5
F	4
Total	34

4

Goals scored	Frequency
0	8
1	7
2	5
3	5
4	2
5	1
Total	28

5

Eggs laid	Frequency
0	2
1	8
2	9
3	11
4	6
5	4
Total	40

6

Fish caught	Frequency
1	2
2	4
3	8
4	10
5	9
6	7
Total	40

7

Form	Frequency
A	5
E	6
L	10
P	5
S	8
W	6
Total	40

8

Price	Frequency
9p	5
10p	7
11p	8
12p	7
13p	8
14p	5
Total	40

9

Minutes late	Frequency
1	10
2	7
3	4
4	6
5	5
6	3
Total	35

10

Pints of milk	Frequency
1	6
2	14
3	8
4	6
5	7
6	4
Total	45

56 BAR CHARTS

Exercise 56A

1 (a) 7 (b) 14 (c) 27
2 (a) 96 (b) 32 (c) Wed. (d) Sat.
3 (a) Mirror (b) 66 (c) 20
4 (a) £4000 (b) Jan. and Apr.
(c) Feb. (d) £4000
5 (a) London (b) Manchester
(c) Dublin and Newcastle (d) 3°C
6 (a) Sat. (b) (i) 18 (ii) 12 (iii) 0
(c) Wed.
7 (a) 25 (b) 40 (c) 85 (d) 180
8 (a) Julie (b) Daniel (c) Martin and Lisa
(d) 25 (e) 40
9 (a) Years 7 & 9 (b) (i) 325 (ii) 210
(c) Year 9 (d) Year 10
10 (a) Jan. (b) Feb.
(c) £85 000 (d) Jan., Mar. and Apr.

Exercise 56B

1 (a) Mon. (b) (i) 22 (ii) 16
(c) Wed. and Thu.
2 (a) Derek (b) Andrew and Tina
(c) Quasim and Ricardo (d) (i) 50 (ii) 45
3 (a) 15 (b) Supermarket (c) 5 (d) 60
4 (a) P (b) A (c) L and W
(d) 305
5 (a) Blue (b) Brown (c) 11 (d) 52
6 (a) 6 (b) Week 5 (c) Week 3 (d) 32
7 (a) Sat. (b) Mon. (c) 0 (d) 25
8 (a) 19–24 (b) 40+ (c) 20 (d) 30
9 (a) Thu. and Fri. (b) Fri.
(c) Mon. (d) 45
10 (a) (i) M (ii) D (b) (i) D (ii) M
(c) S (d) B:26, D:25, Rest:24

Exercises 56C

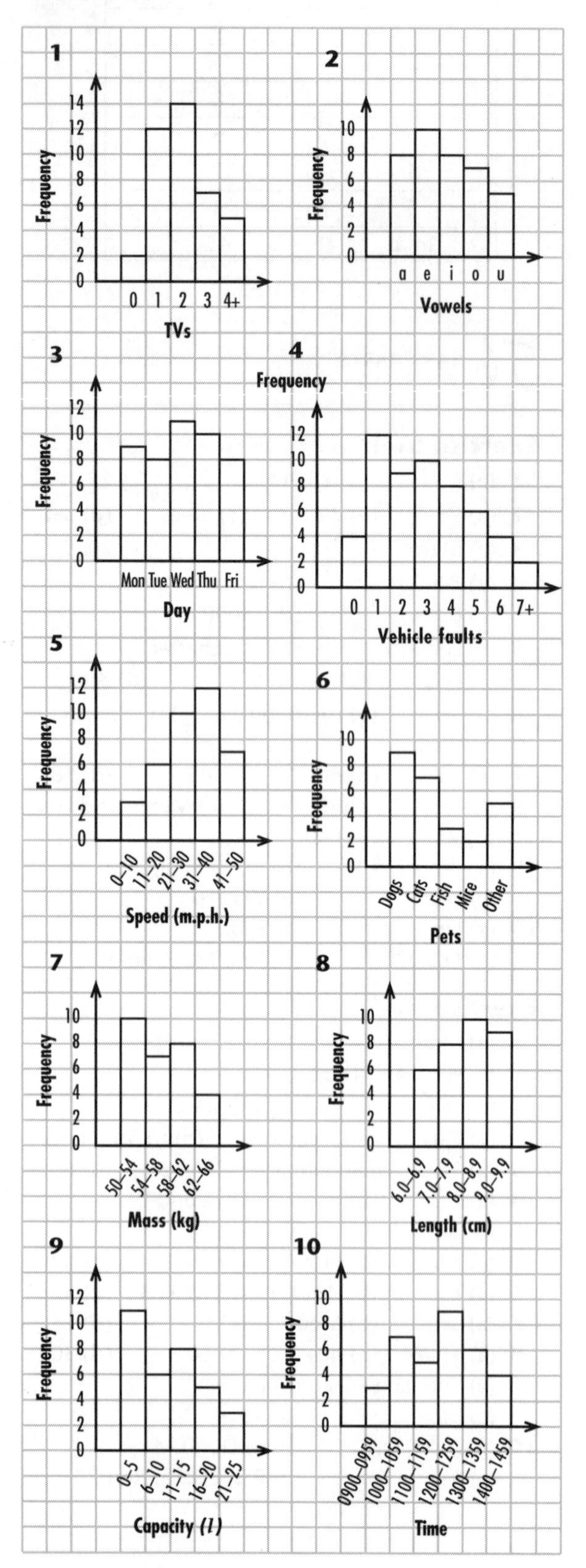

Exercise 56D

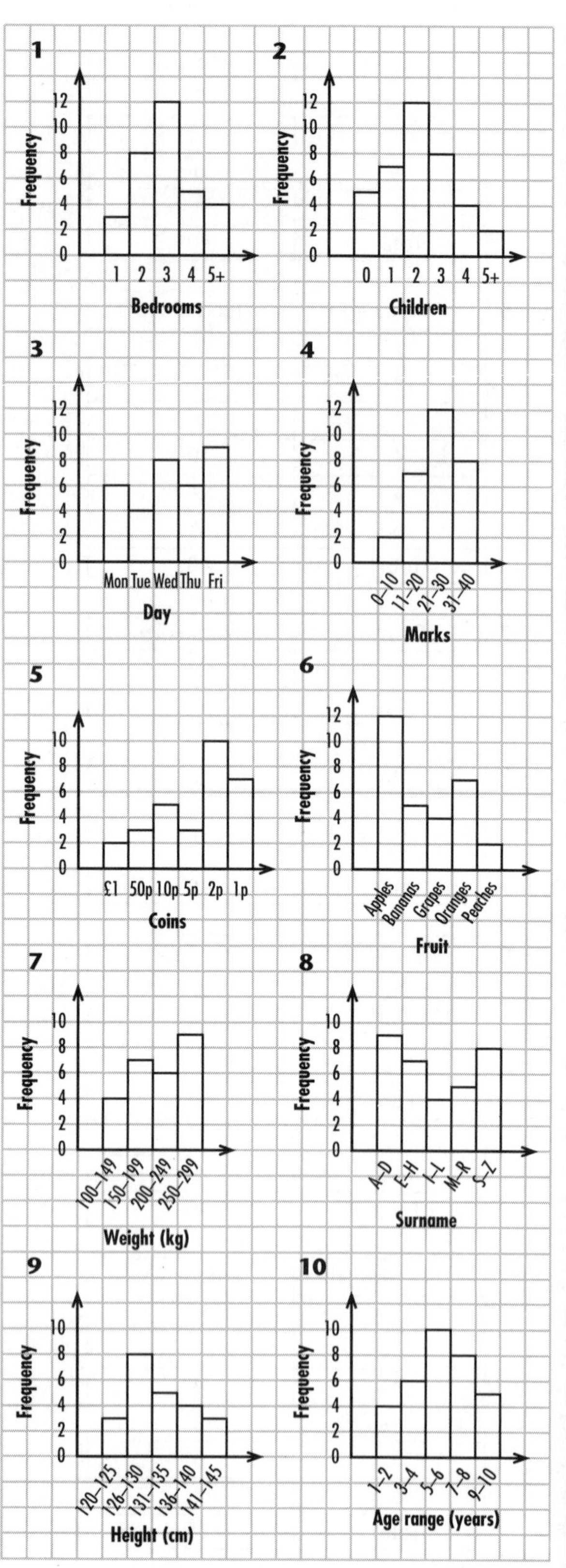

57 PICTOGRAMS

Exercise 57A

1 (a) Strawberry (b) Chocolate chip (c) 6
(d) 21

2 (a) Grapes (b) Bananas (c) 5
(d) 8

3 (a) Cola (b) Crisps and Sweets
(c) 6 (d) 24

4 (a) Toast (b) Fruit (c) 5
(d) 15

5 (a) Meadow Gate (b) Frimby St
(c) 14, 16, 10, 11

6 (a) 40, 25, 35, 20 (b) 120

7 (a) 32, 26, 14, 27, 25 (b) 124

8 (a) Martin (b) Saleem
(c) 12, 9, 11, 7, 10

Exercise 57B

1 (a) Helen (b) Craig (c) 18

2 (a) Fri. (b) Mon. and Thu.
(c) 10 (d) 44

3 (a) CDs (b) Tapes (c) 10 (d) 6

4 (a) Coffee (b) Milk (c) 4 (d) 19

5 (a) Food (b) £30 (c) £70

6 (a) Mon. (b) Wed. (c) 14 (d) 10

7 (a) (i) 50 (ii) 35 (iii) 25
(b) Week 4 (c) 55

8 (a) 18 (b) 27 (c) Walk (d) 72

Exercises 57C

Symbols and scales (questions 5–8) may differ from those shown.

1

Plain	
Cheese & onion	
Beef	
Chicken	
Smokey bacon	

= 1 packet of crisps

2

Monday	
Tuesday	
Wednesday	
Thursday	
Friday	

= 1 book

3

BBC1	
BBC2	
ITV	
Ch4	

= 1 hour of television watched

4

Red	
Blue	
Green	
Yellow	
Brown	

= 1 colour

5

Bill	
Ali	
Remi	
Alan	
Sanchez	

= 4 goals

6

Walk	
Bus	
Car	
Train	
Other	

= 4 people

7

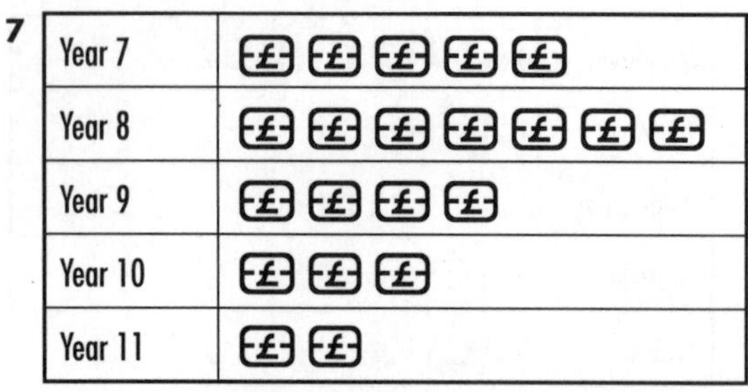

8

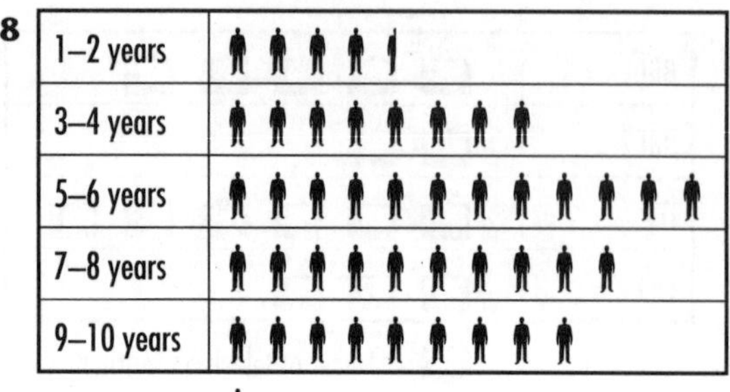

Exercises 57D

Symbols and scales (questions 5–8) may differ from those shown.

1

Monday	
Tuesday	
Wednesday	
Thursday	
Friday	

= 1 hour of sunshine

2

Sleep	
Work	
Eating	
Travel	
Other	

= 1 hour

3

Crime	
Mystery	
Sci-fi	
Romance	
Adventure	

= choice of book

4

UK	
France	
Spain	
USA	
Other	

= holiday destination

5

Monday	
Tuesday	
Wednesday	
Thursday	
Friday	
Saturday	
Sunday	

= 4 birthdays

6

10 a.m.	
11 a.m.	
12 noon	
1 p.m.	
2 p.m.	
3 p.m.	
4 p.m.	

= 50 cars

7

Soul	
Dance	
Rock	
Country	
Heavy metal	

= 4 CDs

8

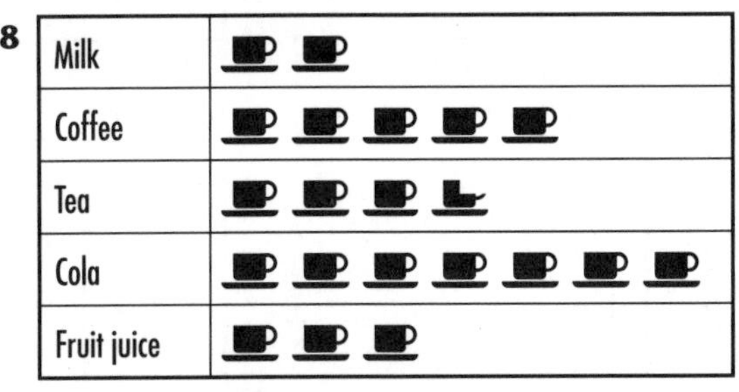

REVISION

Exercise F

1

Number	Frequency
4	6
5	4
6	8
7	9
8	7
9	6
Total	40

2

Letter	Frequency
A	7
B	6
C	4
D	5
E	10
F	8
Total	40

3

Frequency
0 2 4 6 8 10
Mon Tue Wed Thu Fri
Day

4

Frequency
0 2 4 6 8 10 12 14
1 2 3 4 5
Distance (m)

5

Alan	
Ali	
Adam	
Anthony	
Adama	

= goal scored

6

10 a.m.	
11 a.m.	
12 noon	
1 p.m.	
2 p.m.	
3 p.m.	
4 p.m.	
5 p.m.	

= 4 people

Exercise FF

1 (a) 0928 h (b) 0719 h
(c) 1 h 15 min (d) 2 h 8 min
(e) 0559 h (f) 0733 h

2 (a) Mon. (b) Sun. (c) Wed./Fri.
(d) 6 (e) 43

3 (a) Roses (b) Lupins (c) 8
(d) 7 (e) 38

58 INTERPRETING DIAGRAMS

Exercise 58A

1 (a) 7 (b) 5 (c) 3
(d) 4 and 6

2 (a) 10 m (b) 10.5 m (c) 1600 h
(d) 1300–1400 h

3 (a) Week 4 (b) Week 5 (c) 3 mm

4 (a) 1400 h (b) 40°C (c) 39°C
(d) 39.5°C

5 (a) May (b) Jan.
(c) (i) 5 (ii) 4
(d) Feb., Jun., Sep. and Nov.

6 (a) 1940
(b) (i) 4 million (ii) 11 million
(c) (i) 1 million (ii) 2 million

7 (a) 27 (b) 2 p.m.
(c) (i) 23 (ii) 23
(d) (i) 4 p.m. (ii) 10 a.m.

8 (a) (i) 46.4°C (ii) 40.8°C (iii) 42.8°C
(b) (i) 6 min (ii) 3 min (iii) 7 min

Exercise 58B

1 (a) (i) 4 (ii) 5 (iii) 3
(b) (i) 5, 11, 13 and 17 (ii) 9 and 16
(c) 1, 2, 3, 10, 14

2 (a) (i) 9 (ii) 10%
(b) Apr., Mar., Sep.
(c) Aug. and Sep.

3 (a) Clubs (b) Spades (c) 28

4 (a) (i) 14 (ii) 15
(b) (i) 6 and 9 (ii) 4 and 5
(c) 74

5 (a) (i) 9 (ii) 5 (b) Year 11
(c) 30

6 (a) Tue. (b) Fri.
(c) 70, 80, 50, 70, 40

7 (a) 1st week in May
(b) 3rd week in Jul. to 2nd week in Aug.
(c) (i) £520 (ii) £620
(d) (i) 1st week in Jun.
(ii) 4th week in Jun. and 1st week in Jul.

8 (a) (i) 28 (ii) 40
(b) (i) 10 a.m. and 3 p.m. (ii) 2 p.m.
(c) 11 a.m.–12 noon, 2 p.m.–3 p.m.

59 DRAWING BAR LINE GRAPHS

Exercises 59A

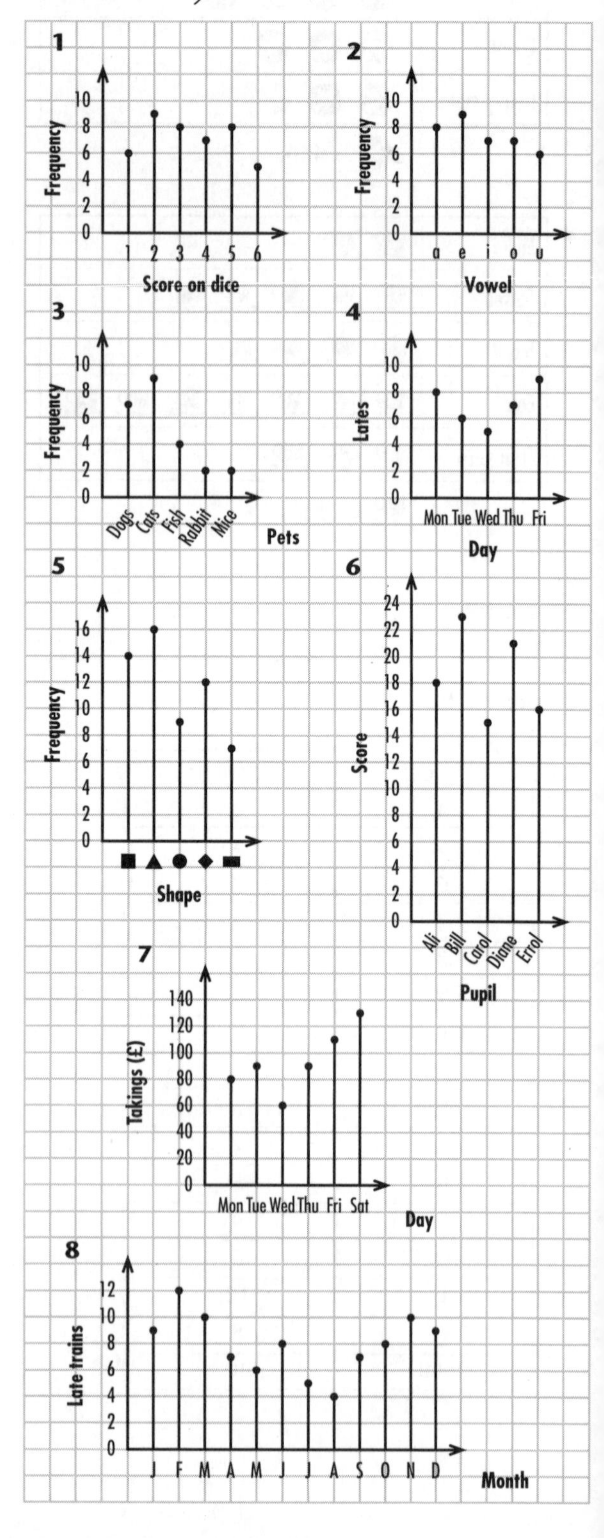

Exercises 59B

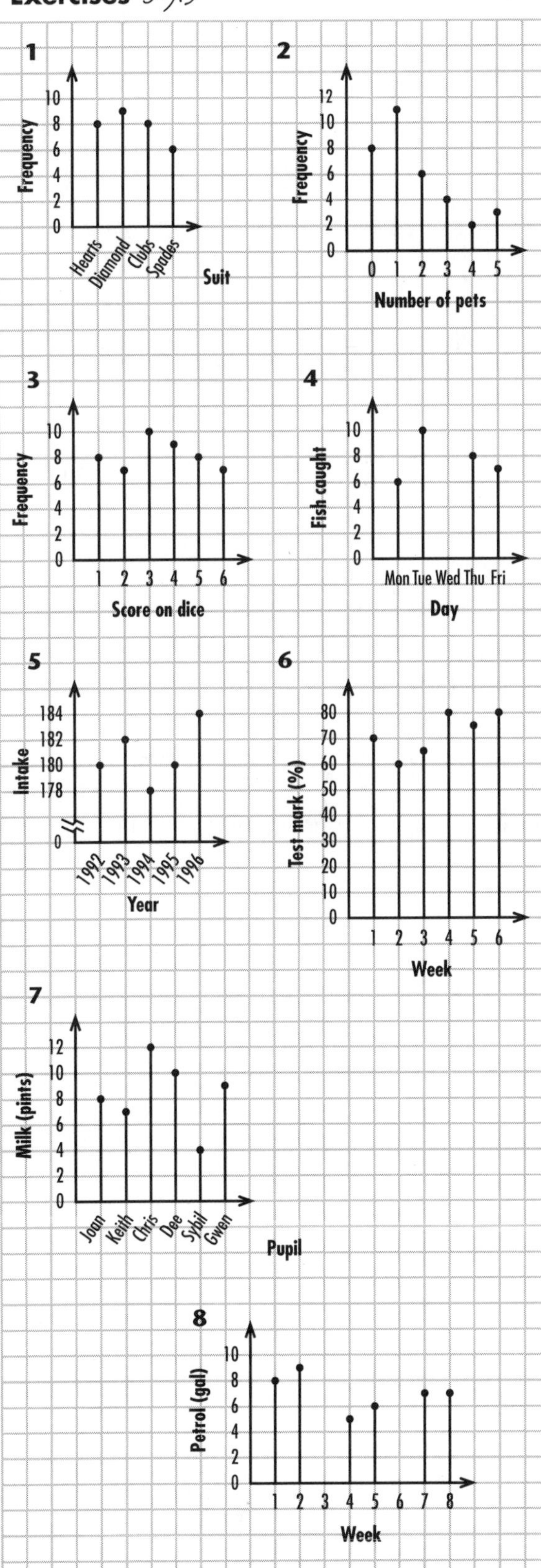

60 DRAWING LINE GRAPHS

Exercises 60A

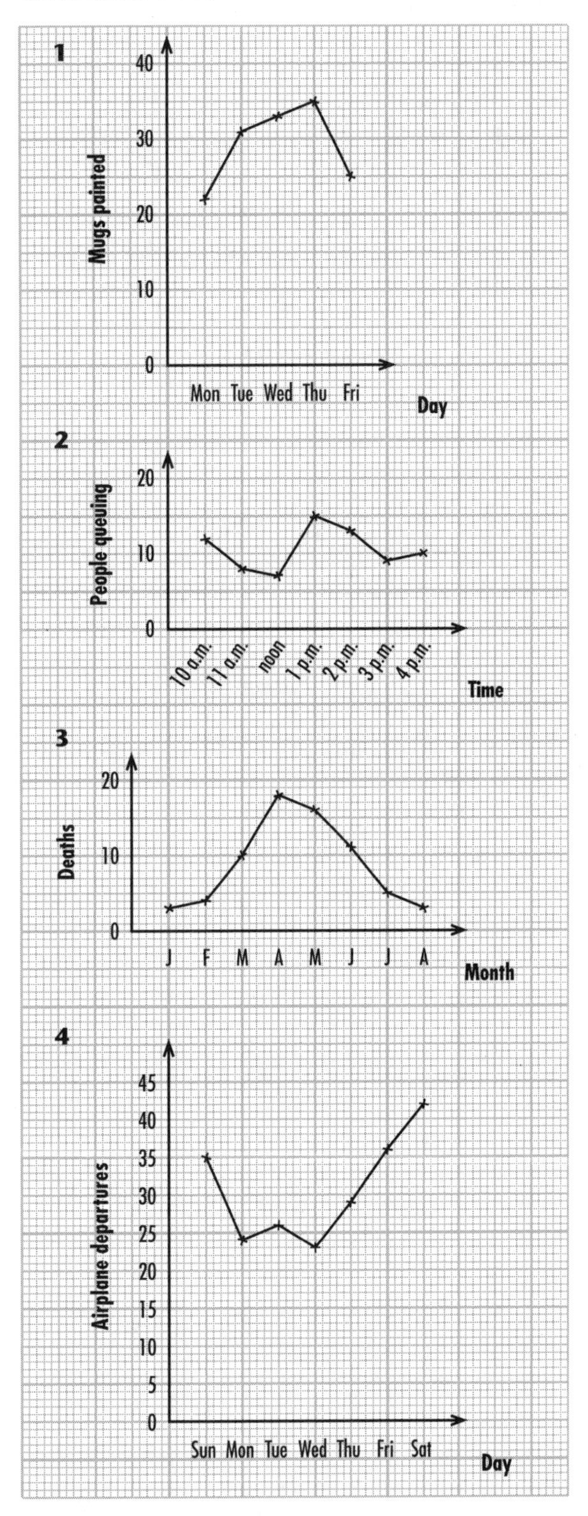

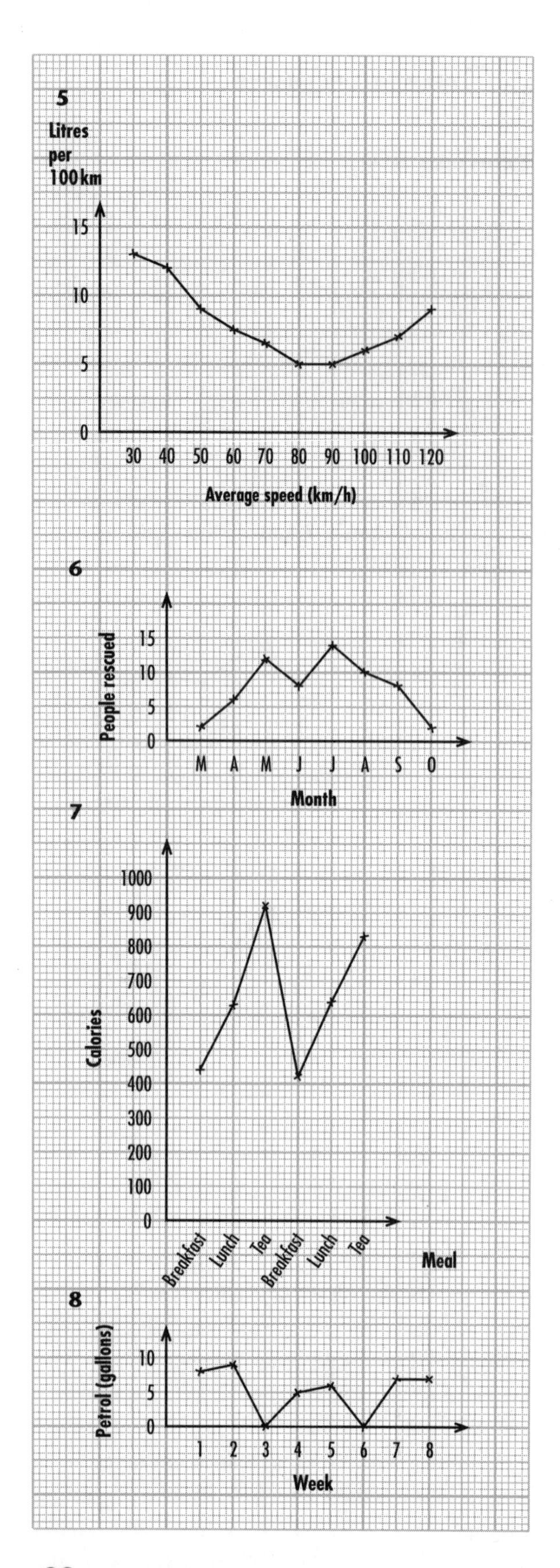

Exercises 60B

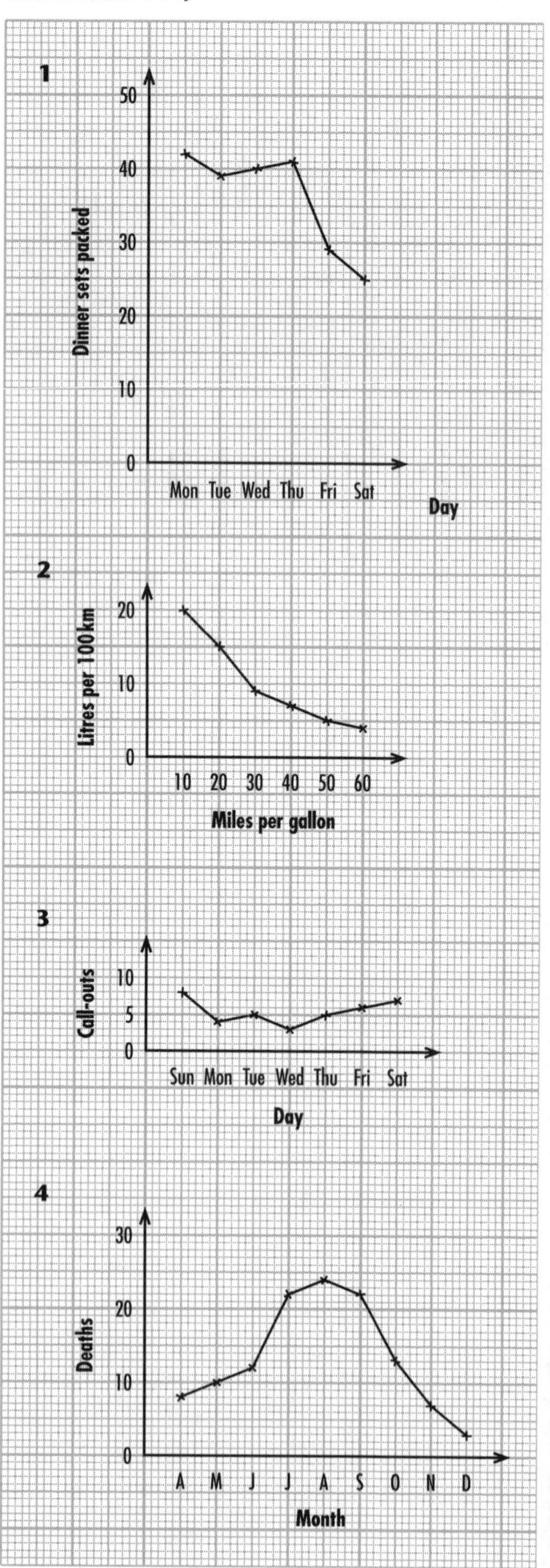

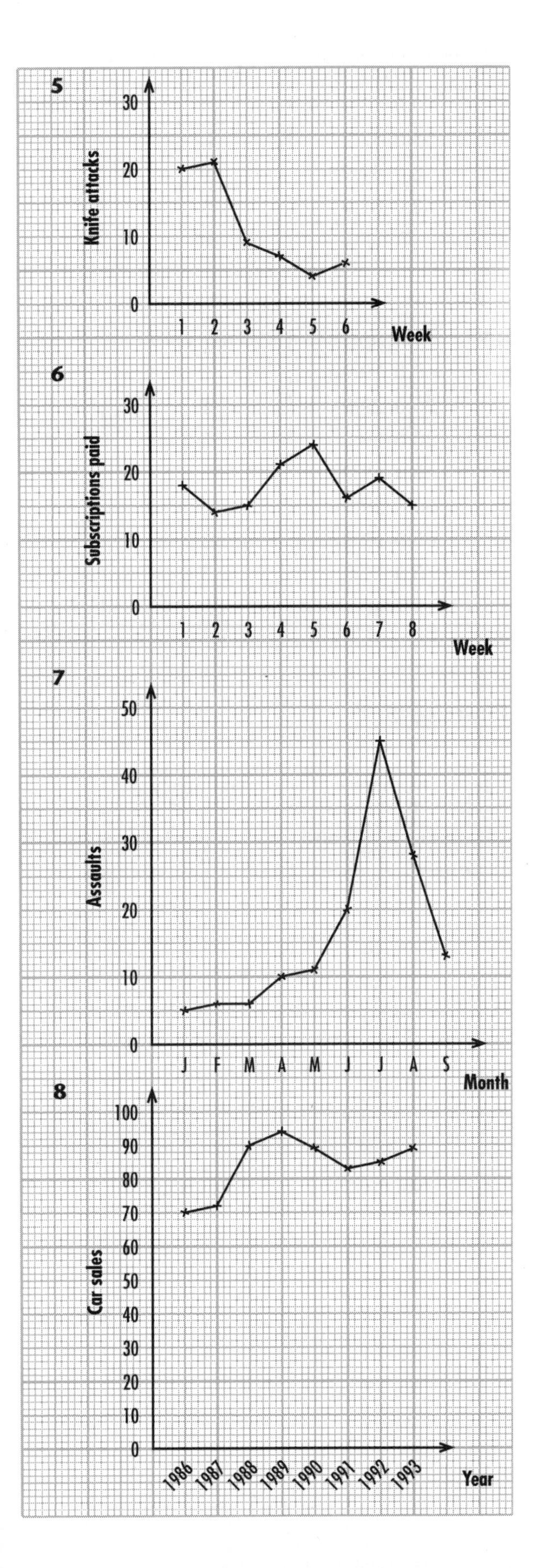

61 GROUPING DATA

Exercise 61A

1

Number	Frequency
0–4	5
5–9	10
10–14	9
15–19	11
Total	35

2

Number	Frequency
30–39	4
40–49	5
50–59	4
60–69	9
70–79	8
80–89	10
Total	40

3

Number	Frequency
10–14	4
15–19	12
20–24	10
25–29	14
Total	40

4

Number	Frequency
20–29	4
30–39	4
40–49	10
50–59	11
60–69	10
70–79	3
80–89	3
Total	45

5

Weights (g)	Frequency
100–149	1
150–199	3
200–249	7
250–299	4
300–349	12
350–399	3
400–449	5
Total	35

6

Amount	Frequency
£0.00–£9.99	1
£10.00–£19.99	10
£20.00–£29.99	7
£30.00–£39.99	5
£40.00–£49.99	2
Total	25

7

Ages	Frequency
0–9	0
10–19	2
20–29	12
30–39	10
40–49	10
50–59	3
60–69	3
Total	40

8

Marks (%)	Frequency
20–29	2
30–39	7
40–49	6
50–59	6
60–69	7
70–79	6
80–89	10
90–99	6
Total	50

9

Points	Frequency
50.0–53.9	4
54.0–57.9	12
58.0–61.9	9
62.0–65.9	5
Total	30

10

Mileage	Frequency
10.0–19.9	3
20.0–29.9	5
30.0–39.0	8
40.0–49.9	7
50.0–59.9	6
60.0–69.9	3
70.0–79.9	3
Total	35

Exercise 61B

1

Number	Frequency
10–19	0
20–29	2
30–39	4
40–49	4
50–59	4
60–69	6
70–79	6
80–89	9
Total	35

2

Number	Frequency
0–4	3
5–9	5
10–14	6
15–19	5
20–24	4
25–29	5
30–34	6
35–39	2
Total	36

3

Number	Frequency
10–19	6
20–29	11
30–39	12
40–49	9
50–59	0
60–69	2
Total	40

4

Number	Frequency
10–14	4
15–19	6
20–24	16
25–29	14
Total	40

5

Amount	Frequency
£0–£4	3
£5–£9	5
£10–£14	7
£15–£19	9
£20–£24	6
£25–£29	5
£30–£34	5
Total	40

6

Videos	Frequency
0–9	2
10–19	6
20–29	5
30–39	8
40–49	9
50–59	12
60–69	8
Total	50

7

Marks (%)	Frequency
30–39	3
40–49	6
50–59	6
60–69	11
70–79	9
80–89	15
Total	50

8

Amount	Frequency
£0.00–£9.99	3
£10.00–£19.99	7
£20.00–£29.99	9
£30.00–£39.99	6
£40.00–£49.99	5
Total	30

9

Weight (g)	Frequency
0.00–0.49	3
0.50–0.99	6
1.00–1.49	5
1.50–1.99	7
2.00–2.49	6
2.50–2.99	3
Total	30

10

Speed	Frequency
0.0–9.9	3
10.0–19.9	3
20.0–29.9	9
30.0–39.9	15
40.0–49.9	5
Total	35

62 MODE AND MEDIAN

Exercise 62A

1	(a) 5	(b) 5	**2**	(a) 1	(b) 3
3	(a) 10	(b) 7	**4**	(a) 7	(b) $5\frac{1}{2}$
5	(a) 4	(b) 5	**6**	(a) 7	(b) 6
7	(a) 51	(b) 52	**8**	(a) 13	(b) 15
9	(a) 8	(b) 6	**10**	(a) 11	(b) 11
11	(a) 12	(b) 19	**12**	(a) 8	(b) $8\frac{1}{2}$
13	(a) 3	(b) 3	**14**	(a) 4	(b) 6
15	(a) 15	(b) $15\frac{1}{2}$			

Exercise 62B

1	(a) 8	(b) 7	**2**	(a) 11	(b) 11
3	(a) 9	(b) 8	**4**	(a) 4	(b) $3\frac{1}{2}$
5	(a) 1	(b) 3	**6**	(a) 24	(b) 28
7	(a) 34	(b) 34	**8**	(a) 23	(b) 21
9	(a) 9	(b) 9	**10**	(a) 13	(b) 14
11	(a) 12	(b) 12	**12**	(a) 16	(b) 15
13	(a) 4	(b) 4	**14**	(a) 9	(b) 8
15	(a) 25	(b) $23\frac{1}{2}$			

63 PROBABILITY SCALE

Exercise 63A

1 I	**2** E	**3** B	**4** C
5 I	**6** H	**7** B	**8** C
9 E	**10** A	**11** F	**12** I
13 B	**14** F	**15** H	**16** E
17 B	**18** A	**19** G	**20** D

Exercise 63B

1 I	**2** E	**3** B	**4** C
5 G	**6** A	**7** D	**8** C
9 B	**10** H	**11** I	**12** D
13 F	**14** B	**15** G	**16** A
17 E	**18** H	**19** I	**20** E

64 OUTCOMES

Exercise 64A

1 1, 2, 3, 4, 5, 6

2 a, b, c, d, e

3 1, 2, 3, 4, 5, 6, 7, 8, 9, 10, 11, 12, 13, 14, 15, 16, 17, 18, 19, 20

4 1, 2, 3, 4, 5, 6, 7, 8, 9, 10, 11, 12

5 HHH, HTT, HTH, TTH, HHT, HTH, THH, TTT

6 1, 2, 3

7 1H, 2H, 3H, 4H, 5H, 6H, 1T, 2T, 3T, 4T, 5T, 6T

8 1A, 1B, 1C, 2A, 2B, 2C, 3A, 3B, 3C

9 AD, AE, AF, BD, BE, BF, CD, CE, CF

10 12, 13, 14, 15, 16, 21, 22, 23, 24, 25, 26, 31, 32, 33, 34, 35, 36, 41, 42, 43, 44, 45, 46, 51, 52, 53, 54, 55, 56, 61, 62, 63, 64, 65, 66

11 Mo/Tu, Mo/We, Mo/Th, Mo/Fr, Mo/Sa, Mo/Su, Tu/Mo, Tu/We, Tu/Th, Tu/Fr, Tu/Sa, Tu/Su, We/Mo, We/Tu, We/Th, We/Fr, We/Sa, We/Su, Th/Mo, Th/Tu, Th/We, Th/Fr, Th/Sa, Th/Su, Fr/Mo, Fr/Tu, Fr/We, Fr/Th, Fr/Sa, Fr/Su, Sa/Mo, Sa/Tu, Sa/We, Sa/Th, Sa/Fr, Sa/Su, Su/Mo, Su/Tu, Su/We, Su/Th, Su/Fr, Su/Sa

12 YYY, BBB, WWW, YBB, BYB, BBY, BWW, BWB, WWB, YWW, WYW, YWW, BBW, BWB, WBB, YYB, YBY, BYY, YYW, YWY, WYY, YBW, YWB, BYW, BWY, WBY, WYB

Exercise 64B

1 H, T

2 1, 2, 3, 4, 5, 6, 7, 8

3 a, e, i, o, u

4 Mon., Tue., Wed., Thu., Fri., Sat., Sun.

5 2, 4, 6, 8, 10, 12, 14

6 A, B, C, D, R

7 11, 12, 13, 14, 15, 16, 17, 18, 21, 22, 23, 24, 25, 26, 27, 28, 31, 32, 33, 34, 35, 36, 37, 38, 41, 42, 43, 44, 45, 46, 47, 48, 51, 52, 53, 54, 55, 56, 57, 58, 61, 62, 63, 64, 65, 66, 67, 68, 71, 72, 73, 74, 75, 76, 77, 78, 81, 82, 83, 84, 85, 86, 87, 88

8 R1, R2, R3, G1, G2, G3, B1, B2, B3

9 A1, A2, A3, A4, A5, B1, B2, B3, B4, B5 C1, C2, C3, C4, C5, D1, D2, D3, D4, D5

10 BBB, RRR, BBR, BRB, RBB, BRR, RRB, RBR

11 RRR, BBB, GGG, RBB, BRB, BBR, BGG, BGB, GGB, RGG, GRG, RGG, BBG, BGB, GBB, RRB, RBR, BRR, RRG, RGR, GRR, RBG, RGB, BRG, BGR, GBR, GRB

12 HHHH, HHTT, HHTH, HTTH, HHHT, HHTH, HTHH, HTTT, THHH, THTT, THTH, TTTH, THHT, THTH, TTHH, TTTT

EVISION

Exercise G

1

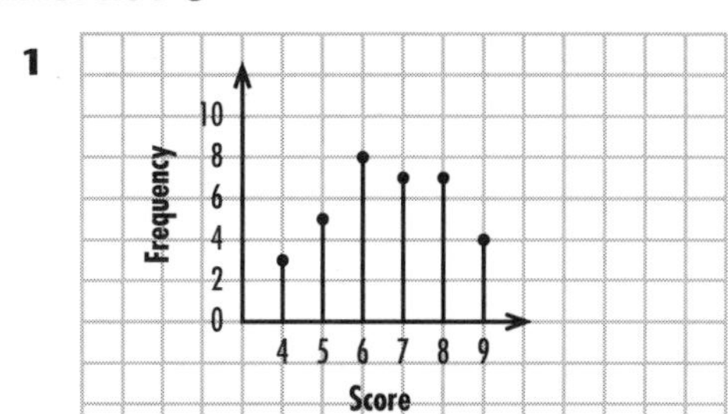

2

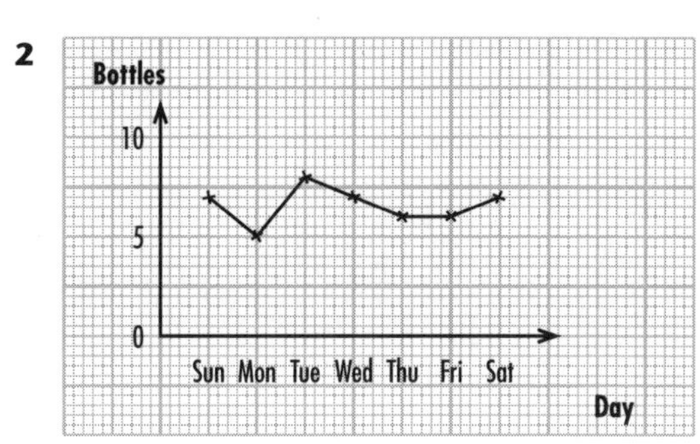

3

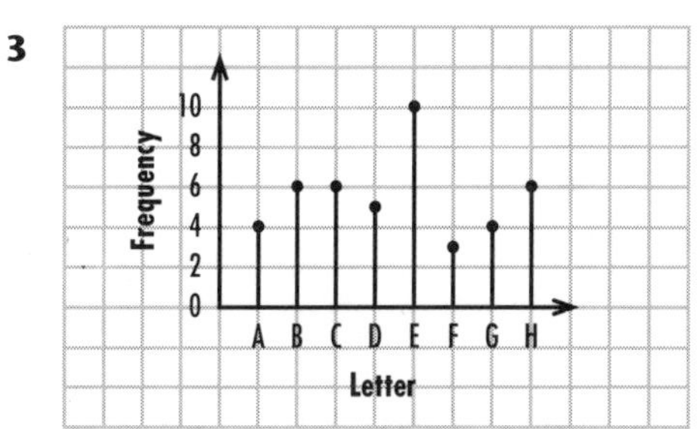

4

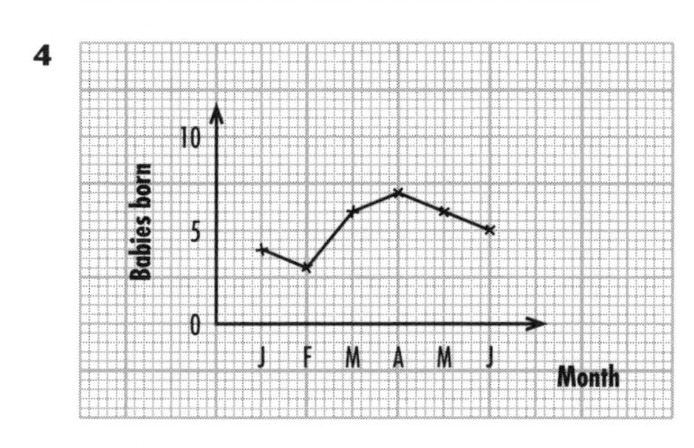

5 (a)

Number	Frequency
0–9	4
10–19	6
20–29	5
30–39	5
40–49	7
50–59	6
60–69	4
Total	37

(b)

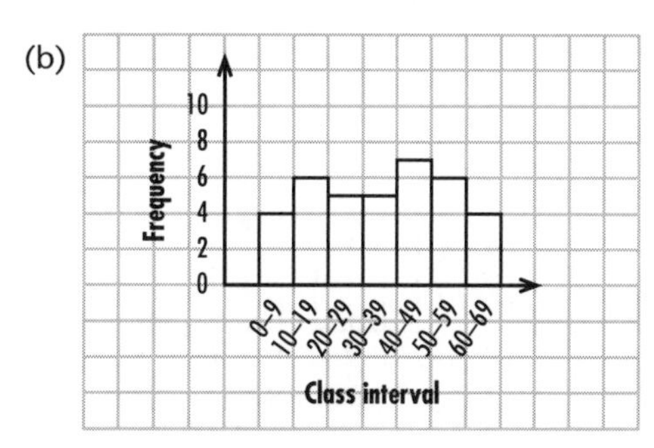

6 (a)

Number	Frequency
0–4	2
5–9	4
10–14	5
15–19	7
20–24	5
25–29	6
30–34	4
Total	33

(b)

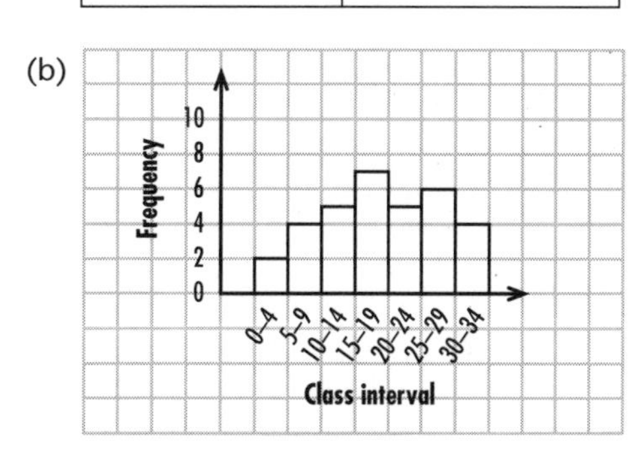

7 (a)(i) 5 (ii) 4 (b)(i) 4 (ii) $3\frac{1}{2}$ (c)(i) 22 (ii) 20 (d)(i) 46 (ii) $45\frac{1}{2}$

Exercise GG

1 (a) Wai-Lee (b) Bill and Remi (c) 5

2 (a) Sat. (b) Thu. (c) 5 (d) 35

3 (a) 12 noon (b) 13°C
(c) 8 p.m. and 12 midnight

4 (a)

Score on dice	Frequency
1	4
2	5
3	4
4	7
5	6
6	5
Total	31

(b)

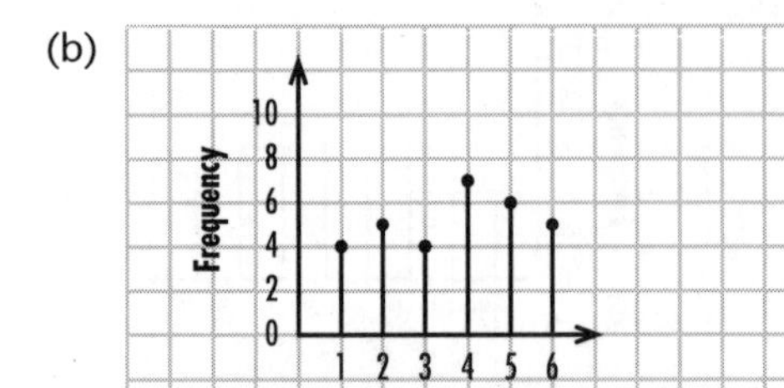

(c) 4

5 (a)

Amount	Frequency
£0.00–£9.99	4
£10.00–£19.99	9
£20.00–£29.99	7
£30.00–£39.99	8
£40.00–£49.99	5
£50.00–£59.99	2
£60.00–£69.99	3
£70.00–£79.99	2
Total	40

(b)

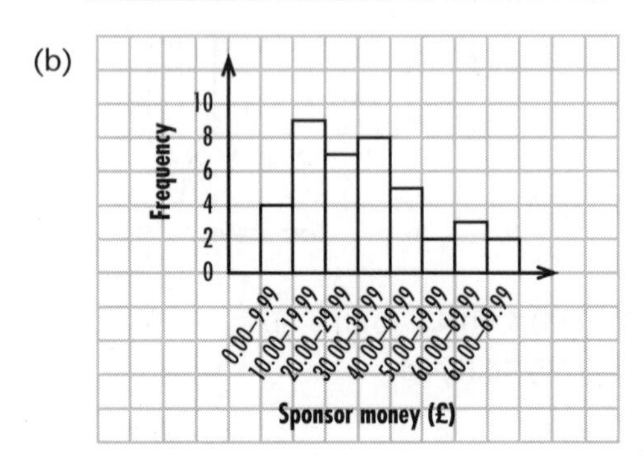

(c) £10.00–£19.99

6 Q, R, S, T, U, V, W, X, Y, Z

7 1HH, 1HT, 1TH, 1TT, 2HH, 2HT, 2TH, 2TT, 3HH, 3HT, 3TH, 3TT, 4HH, 4HT, 4TH, 4TT, 5HH, 5HT, 5TH, 5TT, 6HH, 6HT, 6TH, 6TT

8 (a) A (b) E (c) B (d) I (e) H